# SIKHONA

## THE MAN BEHIND A TECHNOLOGICAL REVOLUTION

MICHAEL THOMPSON

Copyright © 2025, Strive Press LLC, Printed in the United States

All rights reserved. No part of the material protected by this copyright notice may be reproduced or utilized in any form or by any means, electronic or mechanical, including photocopying, recording, or by any information storage and retrieval system, without permission from the copyright owner.

Under no circumstances will any blame or legal responsibility be held against the publisher or author for any damages, reparation, or monetary loss due to the information contained within this book. Either directly or indirectly. You are responsible for your own choices, actions, and results. For related titles and support materials, visit our online catalog at www.strivepress.com

**Legal Notice & Disclaimer Notice:**

This book is copyright-protected. This book is only for personal use. You cannot amend, distribute, sell, use, quote, or paraphrase any part or the content within this book without the consent of the author or publisher.

Please note the information contained within this document is for educational and entertainment purposes only. All effort has been executed to present accurate, up-to-date, reliable, and complete information. No warranties of any kind are declared or implied. Readers acknowledge that the author is not engaging in the rendering of legal, financial, medical, or professional advice.

The content within this book has been derived from various sources. Please consult a licensed professional before attempting any techniques outlined in this book.

By reading this document, the reader agrees that under no circumstances is the author responsible for any losses, direct or indirect, which are incurred as a result of the use of the information contained within this document, including, but not limited to, — errors, omissions, or inaccuracies.

ISBN: 979-8-89725-016-5 *Hardback*

ISBN: 979-8-89725-015-8 *Paperback*

ISBN: 979-8-89725-017-2 *Ebook*

ISBN: 979-8-89725-018-9 *Audiobook*

# Dedication

*To every young, innovative mind from
communities of color—this is for you.
May you dream boldly, create fearlessly, and rise
unapologetically. Let no doubt, no barrier, and no voice of
opposition deter you from building a brighter future for
yourself, your family, and the world.
You are the architects of tomorrow, the visionaries of a new
era. Sikhona stands as a testament that we are here, we are
seen, and we are unstoppable.
Carry the torch, and let your light
inspire generations to come.*

# CONTENTS

| | |
|---|---|
| Foreword | vii |
| Introduction | xi |
| Preface | xv |

1. Early Beginnings – "A Curious Mind Unleashed" — 1
2. Seeds of Innovation - "Creative Youth" — 7
3. Shaping Aspirations - "Early Adolescence" — 11
4. Embracing the Hustle "High School Years" — 15
5. Breaking Boundaries "Young Adulthood" — 19
6. A Visionary Without a Degree – "Breaking the Mold" — 23
7. Visionary on the Rise — 28
8. The First Breakthrough – "Food on the Move" — 31
9. Building a Prototype That Changed It All – "StreamNow" — 42
10. The Big Players Take Notice – "Winning Where Giants Failed" — 48
11. Entrepreneurial Ventures – "Beyond the Prototype" — 57
12. The Birth of Sikhona – "More Than Just a Company" — 63
13. Leading with Purpose – "The Power of Vision and Values" — 72
14. The Future of Technology – "What's Next for Sikhona" — 77
15. A Legacy in the Making – "Building Beyond the Present" — 82
16. Discussion And Reflection Questions — 89
17. Making A Difference Together — 104

# FOREWORD

By Rev. Dr. Albert "Al" Sampson

I have had the honor of walking alongside some of the greatest leaders and visionaries of our time, but few encounters have shaped my life as profoundly as my connection with Michael Thompson. Before I delve into our story, let me first tell you a bit about my own journey.

In 1966, I was one of three men ordained by Dr. Martin Luther King Jr., and I was the only one that was a part of his organization, the Southern Christian Leadership Conference (SCLC). This was a moment that not only defined my ministry but solidified my commitment to justice, equality, and the liberation of our people. Through my work in the Civil Rights Movement and my tireless advocacy for economic empowerment, I have sought to uplift and unite our community. This mission eventually took me to Africa, where I expanded my ministry, laying the groundwork for partnerships that would strengthen ties between the diaspora and the motherland.

It was many years later that I met Michael Thompson. What began as a simple request for an interview turned into one of the most significant relationships of my life. From the moment we spoke, I recognized in Michael a kindred spirit—a man deeply committed to his vision, his people, and the advancement of humanity. Over time, our bond grew, transforming into something akin to a father-son relationship. We speak daily, and in Michael, I have found a trusted confidant and a relentless doer.

Michael is the only person who has truly helped me without ever asking for anything in return. That kind of selflessness is rare, but it is matched by his remarkable ability to follow through. Time and time again, he has fulfilled every promise he made to me, delivering with a precision and integrity that is nothing short of extraordinary.

Michael's work is a continuation of the vision that Dr. King, myself, and many others set forth—a vision of liberation and empowerment. He has taken the brilliance of his mind and combined it with the power of technology to create tools and opportunities that uplift and inspire. There are stories in his journey that all people, especially our young Black boys and girls, should know. His example serves as a beacon, showing them what is possible when intellect, integrity, and a heart for service come together.

One of the many ways Michael has impacted my life is through his technological expertise. When I needed help advancing my nonprofit, Soul Slate PAC, Michael stepped in, using his technological brilliance to transform the way we operated. His solutions were not only practical but visionary, demonstrating his deep understanding of how technology can drive social change.

But Michael's contributions go far beyond the technical. His vision for the world is expansive, and his commitment to Africa and its people is unwavering. He sees possibilities where others see challenges, and his work reflects a profound dedication to building bridges, empowering communities, and creating lasting impact. Michael is a man of integrity—a leader who inspires trust, admiration, and respect.

As you turn the pages of this book, you will discover the depth of Michael's brilliance and the magnitude of his mission. This is not just a story about technology or innovation; it is a story about a man whose life's work embodies the values of service, excellence, and an unyielding commitment to humanity.

Enjoy your read of this brilliant man and his powerful mission.

Rev. Dr. Albert "Al" Sampson

# INTRODUCTION

**"The Architect of Digital Evolution"**

My name is Michael Thompson, and I've always believed in challenging the limits of what people think is possible. Sikhona—a term from South Africa meaning "We are here to be seen"—is more than a name; it's a declaration. For too long, history has tried to erase us, to silence our voices and diminish our presence. But Sikhona stands as a bold affirmation: we are here, we are visible, and we are powerful. This is not just a company; it's a movement rooted in technology and purpose, aimed at uniting the African diaspora and empowering communities globally.

From the very beginning, I envisioned Sikhona as a catalyst for revolution in the digital landscape. Our ten platforms are not just about innovation—they are about creating bridges that span continents, connecting people to opportunities, and enabling generations to build wealth while reclaiming their narratives. Through fiber optics, wireless networks, satellite infrastructures, and cutting-edge technologies, we are not merely participants in the digital age; we are redefining its boundaries. Sikhona's mission is as much about technology as it is about humanity, ensuring that our advancements serve people rather than exploiting them.

Looking back, I can see the seeds of this vision in my own journey. I've always had a knack for seeing what others overlook—a tendency to question, to disrupt, and to reimagine. Whether it was designing groundbreaking network infrastructures or pioneering video streaming before Netflix even knew what streaming was—I was streaming video in 1999 when they were just a two-year-old company focused on DVDs—I have always thrived at the intersection of creativity and innovation. Sikhona represents the culmination of this mindset—a company designed not just to succeed but to leave a legacy of empowerment and transformation.

But make no mistake, this journey hasn't been without its challenges. Over the years, I've faced resistance, skepticism, and countless roadblocks from those who couldn't see the vision. Yet every setback became a stepping stone, every doubt a source of determination. Sikhona's story is a testament to resilience, to the power of dreaming big and daring to act.

This book is not just about me or even about Sikhona—it's about the untapped potential within all of us. It's about recognizing that the same tools that have historically been used to oppress can be wielded to uplift and transform. It's about building not just companies but movements, not just wealth but legacies.

So I invite you to journey with me, to explore the trials and triumphs that shaped Sikhona and the vision that fuels its mission. Together, we'll uncover what it means to be an "Architect of Digital Evolution" and how one man's dream can spark a revolution. This is Sikhona: a beacon of hope, a testament to perseverance, and a reminder to the world that we are here to be seen.

# PREFACE

When I first began this journey of writing, my intention was simple: to help our potential investors understand me—the man behind Sikhona—and the vision driving this company. But as I sat down to put my story into words, it became clear that this was about much more than just me or Sikhona. It was about the lessons, the trials, and the triumphs that shaped my path. It was about creating something that could inspire others—especially the young innovators of today who will carry the torch forward tomorrow.

Our stories matter. For far too long, the narratives of the African diaspora and communities of color have been overlooked, diminished, or told through someone else's lens. This book is my contribution to changing that. As Sikhona's work now brings my journey back to the Motherland, I feel an immense sense of purpose. Our mission is not just about building technology—it's about building bridges, creating opportunities, and making a positive impact for generations to come.

As I write these final words, I am just weeks away from receiving the African Achievement Award as the Male

Entrepreneur of the Year for 2024-2025. It's an honor I don't take lightly. This recognition is a reminder of how far I've come, but more importantly, it's a testament to the power of hard work, vision, and the relentless pursuit of a dream.

This book is written for you—the dreamers, the builders, and the innovators. It's for every young person who has faced doubt, discouragement, or obstacles and kept moving forward. It's for those who see the world not just as it is but as it could be.

I believe in your potential. I believe in your ability to overcome challenges, to create something extraordinary, and to leave a legacy that inspires others. Let this book serve as both a testament to what's possible and a roadmap for what lies ahead.

Thank you for allowing me to share my journey with you. My hope is that these words, these stories, and this legacy will help guide you on your own path—wherever it may lead.

We are here. We are seen. And the future is ours to build.

Michael Thompson

# CHAPTER 1
# EARLY BEGINNINGS – "A CURIOUS MIND UNLEASHED"

I'VE ALWAYS BEEN A DREAMER. Some people grow out of it, but for me, it was a permanent part of who I was—curiosity baked into my DNA. Growing up in Northern Virginia, I didn't have much, but I had enough: enough books to read, enough toys to take apart, and enough imagination to fill the gaps where resources were lacking.

One of my earliest memories of this curiosity was watching replays of the Apollo missions on TV with my father and Star Trek. I couldn't have been more than four years old, but I was mesmerized. I begged my parents for a model of the lunar module and spent hours carefully piecing it together. By the time I was done, I wasn't just playing with a toy—I was imagining myself as part of the mission, dreaming of flying into space and making history.

By the sixth grade, that fascination with space had only grown. I vividly remember sitting in Mr. Brookstein's classroom when the first space shuttle, Columbia, launched. We watched it together as a class, and for the first time, I understood that technology wasn't just about tools—it was about breaking boundaries.

I remembered the name so clearly because I was living in Columbia, Maryland, at the time, which made it feel even more significant to me, beyond the fact that it was one of the first shuttles to take off. That day, I made a promise to myself: I would be part of something that changed the world.

My curiosity wasn't limited to space. I loved taking things apart just to see how they worked. Radios, clocks, you name it—I'd dismantle them, study their insides, and try to put them back together. Sometimes I succeeded. Sometimes I didn't. But every attempt taught me something new.

One of my proudest experiments as a kid was with my Tyco racing car set. I was determined to make my cars faster than anyone else's. I tried everything: cleaning the contacts, tweaking the gears, and even experimenting with steel wool as a conductor. After countless hours of trial and error, I finally figured it out. When I raced my car, it zipped around the track faster than any other. Winning wasn't the point, though. What mattered was the problem-solving, the process of figuring out how to make something better than it was.

These early experiments were more than hobbies—they were the seeds of the innovator I would later become. They taught me to approach every problem with curiosity and to see obstacles as opportunities for growth. Even when resources were scarce, my imagination made up the difference.

In high school, my curiosity found new outlets. I joined the ROTC and discovered my love for aeronautical science. We had an actual airplane in the classroom—a real one—and I spent hours studying it. I dreamed up ideas like variable vector thrusting, a concept that would later find its way into advanced fighter jets. At the time, I didn't realize how significant those ideas were. I was just a teenager with big dreams and a fascination with flight.

But it wasn't just technology that fascinated me. I was also drawn to people—to the way they thought and the problems they faced. I began to see that technology wasn't just about machines; it was about solving human challenges. It was a tool, a means to an end, and I wanted to master it.

Looking back, those early years were a crucible for my creativity. I didn't have a roadmap, but I had an insatiable curiosity and the determination to figure things out on my own. I didn't know it then, but those qualities would become the foundation of everything I'd go on to build.

By the time I graduated, I was ready to take on the world. I didn't have all the answers, but I had the one thing that mattered most: a belief that anything was possible if you worked hard enough and dreamed big enough.

---

My earliest memories are of curiosity and wonder. I wasn't the type of kid who just accepted the world as it was—I wanted to know how it worked, why it worked, and how I could make it better. I can still picture myself sitting at the window in our family home, watching the neighborhood kids run to catch the MTA (Mass Transit Administration) bus every morning. They'd be laughing, shouting, and sometimes barely making it on time. I wanted so badly to be one of them. I'd turn to my mother and ask, "When is it my turn to go?"

"Mom, when can I go to school?" I must've asked her that question a hundred times. She'd always give me the same answer: "Soon, Michael. You're not old enough yet." But "soon" wasn't soon enough for me. I was eager to learn, to be part of something bigger than what I already knew. To me, school wasn't just a place—it was a gateway to understanding.

Even before I could go to school, I was already developing a reputation in the family. My uncle Wade called me "Mike the Knife" because, as he put it, I was sharp. That nickname stuck, and even as a little kid, I wore it like a badge of honor. It wasn't just a nickname—it was an acknowledgment of my curiosity and intelligence. I wanted to prove, even at that young age, that I could figure things out.

One of the first things I remember building was a model of the lunar module. I couldn't have been more than three or four years old, but I remember sitting on the floor, carefully assembling the pieces. It came with tiny astronauts, and I'd place them on the little model moon surface, imagining what it would be like to stand there myself. The whole process fascinated me. It wasn't just a toy—it was a project, something I could create with my own two hands.

That curiosity didn't stop with toys. I wanted to understand everything around me. I was the kind of kid who'd take apart my Christmas toys—sometimes using my dad's tools when I shouldn't have—to see what was inside. Sometimes I could put them back together; sometimes I couldn't. But that didn't matter to me—I was learning. My parents probably weren't thrilled about me dismantling my toys or sneaking my dad's tools, but they let me explore because they could see how much I loved it.

One of my favorite things to do was experiment with my toys. I had an electric race car set that I absolutely loved. Most kids would have just raced the cars and been done with it, but not me. I wanted those cars to go faster. So I started experimenting. I opened up the motors, swapped out the copper brushes for steel wool, and suddenly, my cars were unstoppable. My friends couldn't figure out how my cars always won. I didn't tell them my secret—I just let them think I had some kind of magic touch.

Then there were the seeds. My mom used to buy watermelons in the summer, and I'd save the seeds. I didn't want to just throw them away—I wanted to see if I could make them grow. I found an old can from her cheese dip, filled it with dirt, and planted the seeds. For sunlight, I rigged up my Light Bright to mimic the sun, using its single incandescent bulb as the source of light. Every morning, I'd check on those seeds, waiting for the first little green sprout to appear. When it finally did, I felt like I'd unlocked some kind of secret to life. It wasn't just about growing a plant—it was about proving to myself that I could make something happen.

My brother Eric and I were partners in almost everything. We shared a room, which meant we shared ideas, experiments, and plenty of laughs. We'd build go-karts out of scrap wood, tape, and whatever else we could find. I wasn't satisfied with just pushing the kart around—I wanted to add a motor. I'd take the tiny motors from old toys and try to rig them up to make the kart move on its own. Most of the time, it didn't work, but every now and then, I'd get it just right. Those moments of success were pure magic.

Even my playtime was about building and creating. I loved folding paper airplanes, trying out different designs to see which ones would fly the farthest. I'd spend hours tweaking the folds, adjusting the angles, and testing them out in the backyard. It wasn't just about having fun—it was about solving problems and figuring out what worked.

Of course, my curiosity wasn't always easy for my parents to manage. They'd come home to find I'd taken my stuff apart. My mom would shake her head and say, "Michael, why can't you just leave things alone?" But she always said it with a little smile, like she knew I couldn't help myself.

One Christmas, my parents gave me a set of walkie-talkies. I was thrilled. But instead of just using them to talk to my brother from across the house, I wanted to know how they worked. Within a week, I'd taken them apart and was trying to turn them into some kind of intercom system. It didn't quite work, but I learned more in that week than I could've in a month of just playing with them.

Looking back, those early years were all about exploration and discovery. I didn't have much, but I made the most of what I had. Whether it was building a motorized boat out of toy parts or figuring out how to make my race cars faster, I was always chasing the thrill of creating something new. My family might not have understood my endless tinkering, but they supported me in their own way. They saw that spark in me and let me run with it.

What I didn't realize at the time was how much those moments were shaping me. Every toy I took apart, every experiment I tried, and every seed I planted was teaching me to think critically, to solve problems, and to dream big. Those lessons would become the foundation for everything I'd do later in life. Even now, I look back at those early years and smile. That little boy sitting by the window, watching the world go by, had no idea how much he'd grow, how far he'd go, or how much he'd accomplish. But he was ready for the journey.

# CHAPTER 2
# SEEDS OF INNOVATION – "CREATIVE YOUTH"

WHEN YOU'RE a kid with a big imagination and no money, you learn how to get creative. Birthdays in our house were simple, but I wanted to make them special for my parents. The problem was, I didn't have the cash to buy gifts, so I did the next best thing—I made them. I'd take apart my toys, the ones I loved, and turn them into gifts. Imagine an 8-year-old handing over a homemade contraption pieced together from race cars and action figures, wrapped in love and tape. It wasn't fancy, but it was mine, and I think they appreciated the thought.

That need to create, to do something meaningful, was in everything I did. I didn't just see things for what they were—I saw what they could be. I had a knack for looking at the ordinary and imagining something extraordinary. Like the watermelon seeds from my mom's kitchen. Most kids might spit them out and move on, but not me. I'd scoop up a few, grab an empty can from the cheese dip she liked, and fill it with dirt. Then I'd plant those seeds and set up my very own "growth lab" using a Light Bright as an artificial sun.

The incandescent light bulb was my version of sunlight, and in my mind, it was just as effective as the real thing.

I'd water the seeds, turn on the Light Bright, and watch. To me, it wasn't just about growing a plant; it was about creating life, seeing something go from an idea to a reality. I'd fall asleep next to my little experiment and wake up to see tiny green shoots breaking through the dirt. It was magic, pure and simple, and I was hooked. The thrill of seeing something I'd planted take root and grow was like discovering a secret about how the world worked.

It wasn't just plants or toys either. I loved tinkering with anything I could get my hands on. My parents got me a children's guitar when I was four, and I spent hours plucking away at the strings. Music was this whole other language to me, and even as a kid, I wanted to figure it out. I wasn't content with just strumming—I wanted to understand how sound worked, why some notes felt happy and others felt sad. And then there was my fascination with insects. I'd catch them, study them, and try to understand how something so small could be so complex. I guess you could say I was always chasing knowledge in one form or another.

One of my favorite projects was building a motorized boat. I'd take the motors out of toy cars and find ways to attach them to a cheap plastic boat. I didn't have fancy tools or materials, but I had tape, Elmer's glue, and determination. I'd balance the boat just right, add a propeller, and test it out in the bathtub or the pond down the street. Seeing that boat zip through the water, powered by something I built, felt like a small victory every time.

Even my playtime wasn't ordinary. While most kids were happy with store-bought toys, I was out in the yard building go-karts out of scrap wood or crafting slingshots with rubber bands. I didn't just want to play; I wanted to make the game itself. I'd fold paper airplanes, testing which designs flew farther or faster, always looking for ways to improve. It wasn't about having the best toys—it was about creating something better than what I started with.

I think part of my drive to create came from my surroundings. Growing up, we didn't have a lot of money, but my parents made sure we had enough. Still, I knew better than to ask for expensive toys or gadgets. If I wanted something, I had to figure out how to make it myself. That mindset wasn't just practical—it was empowering. It taught me that I didn't have to rely on others to make things happen. I could take what I had and turn it into something incredible.

I was always experimenting, always learning. One summer, I decided to try building a makeshift telescope. I'd read about how lenses worked in a science book, and I was convinced I could make one out of old magnifying glasses and cardboard tubes. It wasn't perfect—I couldn't see the stars clearly—but the fact that I could get it to work at all was enough to keep me going. I wasn't just interested in science—I was living it.

Another time, I got it in my head to build a small windmill to generate electricity. I didn't have the materials to make it work, but that didn't stop me from trying. I'd draw up plans, test different designs, and imagine how I'd one day build something big enough to power the whole house. My parents didn't always understand what I was doing, but they let me have my space to figure it out.

Looking back, I realize those moments of creativity weren't just about having fun. They were about problem-solving, about learning to look at the world differently. I didn't know it at the time, but those experiments and projects were laying the groundwork for everything I'd do later in life. They taught me to think critically, to ask "what if," and to never settle for what's handed to you. If you want something to be better, you have to make it better.

I spent hours in our apartment and outside, building and testing my creations. My curiosity fueled endless experiments as I dreamed up ideas that seemed impossible. The go-karts I built weren't high-tech—they looked like something straight out of *The Little Rascals*. But while I never added electric motors to those, I did tinker with toys that weren't originally designed to move. I'd add motors to them, making them do things they weren't meant to do. It wasn't just about having fun; it was about seeing what was possible and pushing the limits of my imagination.

As I got older, my projects became more ambitious. I wasn't content to just build for fun—I wanted to see how far I could push myself. I started keeping a notebook where I'd sketch out ideas, jot down experiments, and plan future projects. It wasn't just a hobby—it was a way of life. And even though most of my creations didn't turn out the way I imagined, every failure taught me something new.

Looking back, those years were about more than just building things—they were about building myself. Every project, every experiment, and every challenge helped me grow. They taught me to think outside the box, to embrace failure as part of the process, and to believe in my ability to create something from nothing. Those lessons have stayed with me, shaping not just my career but my entire outlook on life. And for that, I'll always be grateful for the little boy with big dreams and a toolbox full of possibilities.

# CHAPTER 3
# SHAPING ASPIRATIONS – "EARLY ADOLESCENCE"

BY THE TIME I hit middle school, life had started to shift. Things were a little more serious, a little more complicated. My carefree childhood was giving way to a world where responsibilities, emotions, and ambitions began to take shape. Middle school wasn't just about going to class or hanging out with friends—it was about discovering who I was and who I wanted to become.

We lived in Reston, Virginia, a suburb with a mix of everything—diverse neighborhoods, rolling parks, and bustling strip malls. Our family's economic status was what you might call middle-class, but we were on the modest end of that spectrum. We had enough to get by, but not much more than that. My dad worked hard, holding down a steady job, and my mom stretched every dollar to make sure we had what we needed. There weren't extravagant vacations or expensive toys, but there was always food on the table and clothes on our backs. For that, I was grateful.

Our neighborhood wasn't flashy, but it was full of character. We lived in an apartment community, so there weren't any driveways for kids to shoot hoops in, but that didn't stop us. We rode our bikes along the nearby bike path, played basketball at the elementary school next door, and hung out at the Reston Community Center or Tall Oaks Shopping Center.

It was a place where you felt like you belonged, even if things weren't always perfect. I had friends from all kinds of backgrounds—some whose families were better off and others who struggled to make ends meet. That mix gave me perspective and helped shape how I saw the world. Growing up there wasn't just about having fun; it was about understanding what community really meant and appreciating the richness that comes from different experiences.

One of the toughest moments of my early adolescence came during those middle school years. I was sitting in class on an icy winter day when the principal called me to the office for early dismissal. I remember the snow falling outside, and as I walked down the hallway, a knot formed in my stomach. Somehow, I just knew something bad had happened. My mom was waiting for me in the office, her face a mix of sadness and strength. That's when she told me my grandfather—her father—had passed away.

He was more than just a grandfather to me—he was a force of nature, the kind of man who lit up a room and always made time for his grandkids. While he didn't chase us around the yard, he had no problem chasing us through the house when we got too noisy. His ears had been sensitive ever since his time in the Navy during World War II, and though he'd fuss and holler out of love, he never seemed in a real rush to catch us. I think he just enjoyed seeing us run, our laughter echoing through the halls.

He'd tell us the best stories, the kind that made the past come alive, and his hugs had a magic way of making everything feel okay. Losing him was like losing a part of my foundation. It was the first time I truly understood how fragile life could be, and it hit me harder than I ever expected. Grief has a way of creeping in, even when you think you're fine. It took a long time to fully process his absence, but in that journey, I came to realize just how much he had taught me about the importance of family and love—even in the little things, like the way he pretended to chase us, or the warmth of his booming laugh.

While I was dealing with that loss, I was also starting to figure out what I wanted in life. It wasn't always easy to reconcile my dreams with the reality of our family's situation. My dad was a practical man. He believed in what he called the "strong 40"—working 40 hours a week for someone else, earning a steady paycheck, and living a secure life. To him, that was the smart, safe way to live. And for a lot of people, it is. But to me, it felt limiting. I didn't want to just get by—I wanted to build something of my own.

That entrepreneurial spirit wasn't something my family fully understood. I'd talk about ideas I had, businesses I wanted to start, or projects I wanted to take on, and I'd get mixed reactions. My mom would smile and nod, not really sure what to say. My dad, on the other hand, was more direct. He'd tell me to focus on school, get a good job, and "play it safe." We had more than a few heated discussions about it, especially as I got older. He wanted stability for me, while I was chasing possibility.

That tension came to a head when I was nearing the end of high school. I'd been saving up money from odd jobs and small ventures, and I wanted to invest it in one of my ideas. My dad thought I was crazy. "Why risk what you've worked so hard for?" he'd ask. But to me, the bigger risk was not trying at all. We butted heads a lot over that, but I think deep down, he respected my determination, even if he didn't always agree with it.

Middle school was also when I started to notice how different people's paths could be. I had friends I'd grown up with, kids who used to play tag in the street or ride bikes with me around the neighborhood. But as we got older, some of them seemed to get stuck. They didn't have the same drive or curiosity that I did, and it started to show. Some of them got into trouble, others just seemed content to coast through life. It wasn't about judging them—it was about recognizing that I was on a different path.

That realization wasn't easy. I valued those friendships, but I was starting to see that not everyone could—or should—follow me on my journey. It was a hard lesson to learn, but an important one. Sometimes, to grow, you have to let go of what's comfortable. I didn't know it at the time, but those moments were shaping the way I'd approach relationships, work, and life in general.

Looking back, my early adolescence was a turning point. It was a time of loss, discovery, and ambition. I was learning what mattered to me—family, creativity, and the drive to build something meaningful. Those years weren't always easy, but they set the foundation for everything that came next. And even though I didn't always know where I was headed, I knew one thing for sure: I wasn't going to settle for anything less than the life I imagined for myself.

# CHAPTER 4
# EMBRACING THE HUSTLE
# "HIGH SCHOOL YEARS"

HIGH SCHOOL WASN'T JUST about discovering who I was—it was about making things happen. Herndon High School gave me the structure and space to figure out my interests, but it was outside the classroom where I truly found my rhythm. Music wasn't just something I loved—it became my life. And it wasn't long before I found myself not just performing but leading a movement.

It started with my brother Eric and me. We'd been playing around with beats and rhymes for years, but high school gave us the confidence to take it seriously. That's when we formed our rap group. Darryl, better known as DK Jay, was the first to join. He had talent and charisma, but what really stood out was his hunger. Then came Brad, another natural performer. What made our group special was the chemistry we had—not just on stage but in the way we worked together.

Funny thing was, I was the second youngest in the group. Darryl and Brad were both a few years older, but they still deferred to me as the leader. They saw my vision, my drive, and the way I could organize and make things happen. Leadership wasn't something I asked for—it was something I earned. Whether it was writing songs, planning performances, or making decisions about the group's direction, they trusted me to take the lead.

One of the first things I did as our leader was arrange for studio time. Back then, recording wasn't cheap. There was no fancy home studio setup with digital software—everything had to be done in a professional space. I'd hustle to save money, sometimes doing odd jobs or flipping items to make ends meet, just to make sure we could afford to record. I didn't mind, though. To me, investing in the group was worth every penny.

Recording sessions were intense. I'd book the studio and come prepared with a plan. We weren't there to waste time or money. I'd make sure everyone knew their parts, from verses to hooks, before we even stepped into the booth. Once we started recording, I'd keep an eye on the clock and the budget, making sure we got the most out of every session. Those recordings were more than just tracks—they were proof of what we could do when we worked together.

But music wasn't just about what happened in the studio—it was about what happened on stage. That's where my knack for organizing came into play. I didn't want to wait for opportunities to perform; I wanted to create them. That's when I started arranging parties. And these weren't just any parties—they were events.

I'd scout out venues like community centers, school gyms, or even private halls. Securing a location wasn't easy, but I knew how to negotiate. I'd sit down with venue managers, explain my vision, and make sure they knew I was serious. And I didn't stop there. I wanted everything to be above board, so I hired off-duty police officers for security. It was a bold move for a teenager, but I wasn't taking any chances with safety.

I also made sure to involve my family. My dad ran the check-in table at the door, keeping everything organized and ensuring the money was accounted for. My Uncle Gary was in charge of the concession stand. He was great with people, and having family handle the operations gave me peace of mind. My brother Eric was my right-hand man, helping with setup, music, and anything else that needed to be done.

Marketing these events was another challenge, but I thrived on it. Back then, there was no social media to spread the word, so I relied on flyers and word of mouth. I'd design the flyers myself, spending hours making sure they looked professional. Then I'd hit the streets, handing them out at school, in neighborhoods, and even outside local businesses. By the time party night rolled around, the buzz was undeniable.

The night of the event was always electric. I'd arrive early, making sure everything was in place. I'd check in with the police officers, confirm the setup with my dad and uncle, and walk the venue to make sure it looked just right. When the doors opened, the energy was unreal. People would line up outside, waiting to get in, and once they were inside, it was game on.

Performing was the highlight of the night. Our group would hit the stage, and the crowd would go wild. We'd perform the songs we had worked so hard on in the studio, feeding off the audience's energy. But even while performing, I was in leader mode. I'd glance at the security team, check in with my brother or uncle between sets, and keep an eye on the overall flow of the event.

There was one night that stands out above the rest. It

was one of our biggest parties yet. We went all out, bringing in upgraded sound equipment, a light show, and even a VIP section for our biggest supporters. I remember standing on stage, looking out at the crowd, and feeling like we were part of something special. It wasn't just a party—it was a movement, and we were at the center of it.

Of course, there were challenges. Equipment would fail, people would show up late, or last-minute issues would pop up. But I learned how to adapt, how to stay calm under pressure, and how to make quick decisions. If something went wrong, I fixed it. If someone needed guidance, I gave it. Those moments taught me the value of leadership, not just in music but in life.

By the time high school ended, we had made a name for ourselves. We weren't just a rap group—we were organizers, entrepreneurs, and leaders. Those years weren't just about music or parties—they were about learning how to build something from the ground up. And even though we all eventually went our separate ways, the lessons I learned during that time have stayed with me.

Looking back, I see those years as the foundation of everything I've done since. They taught me how to lead, how to take risks, and how to turn dreams into reality. High school wasn't just a chapter in my life—it was the start of a story that's still being written.

# CHAPTER 5
# BREAKING BOUNDARIES
# "YOUNG ADULTHOOD"

BY THE TIME I turned 18, I had a clear vision of the life I wanted to build. Moving out of my parents' house and into my first apartment felt like crossing a threshold into adulthood. I landed a three-bedroom unit at the Fairway Apartments in Reston, Virginia. At first, it seemed excessive for just me, but the complex was near a golf course, and the name "Fairway" gave it a sense of sophistication. Eventually, a friend moved in, and we split the rent. It wasn't far from my parents' home, but it symbolized my independence and determination to carve my own path.

---

**Two Worlds, One Lesson**

Growing up in Reston, a town that straddled two economic worlds, was a lesson in contrasts. On one side were affluent families with sprawling homes, luxury cars, and access to the latest technology. On the other side were middle- and lower-income families, like mine, who made do with less. I had friends from both sides of the tracks, which gave me a unique perspective on opportunity and success.

It didn't take long for me to notice a pattern among the families who seemed to "have it all." Their parents were often entrepreneurs or high-level executives. Some were government officials, while others were professional athletes. The common thread? They didn't settle for ordinary jobs. They built something—businesses, careers, legacies. Observing this as a teenager sparked something inside me. I realized I didn't want to just work a "strong 40," as my dad would call it. I wanted to create something of my own, to be the kind of person who didn't just earn a paycheck but built wealth and opportunity.

---

### Ms. Rosenthal's Surprise

My journey into technology wasn't without its hurdles. In ninth grade, I enrolled in a computer science class taught by Ms. Rosenthal, who was passionate about her subject but had a rigid teaching style that didn't click with me. Concepts felt foreign, and her approach left little room for creative thinking. Frustrated, I eventually dropped the class, deciding I'd figure out computers on my own.

Years later, in 2001, I crossed paths with Ms. Rosenthal again. By then, I was thriving in the tech world, working as a vice president at Verizon and designing groundbreaking network infrastructures. When I told her about my career, her reaction was priceless. "You're in tech now? I never would've guessed!" she exclaimed. I smiled and said, "Yep, but it wasn't your teaching style that got me here." We both laughed, and she admitted that traditional teaching methods don't work for everyone. That moment was a reminder that sometimes, the best lessons aren't taught—they're learned through experience.

---

### Self-Taught Hustle

Back then, having a computer at home was unheard of—no one I knew had one. Libraries didn't even have computers yet, but that didn't stop me from pursuing my interest. I saved up the money I earned from a summer job and bought my first computer around 1981 or 1982. It was a game-changer, opening up a world of possibilities. I taught myself BASIC programming, diving into coding, networking, and troubleshooting. Those early skills became the foundation for my career. My first jobs at companies like AT&T and Quantum Computer Services (which later became AOL) gave me invaluable real-world experience, and I thrived in the fast-paced, ever-evolving world of technology.

I worked tirelessly, putting in eight-hour days at the office and spending another six to eight hours experimenting at home. Weekends were no different. My relentless drive to learn and innovate set me apart, earning me roles that many with traditional degrees couldn't secure. By 19, I was making as much money as adults with families, and by my early twenties, I was earning more than many college graduates.

---

### A Vision of the Future

Even then, I could see where technology was headed. The internet was starting to transform the world, and I believed that phones and the internet would one day merge into indispensable tools. While my peers were focused on immediate successes, I was thinking ahead, positioning myself for a future that most couldn't yet imagine. My parents didn't always understand my choices—especially my decision to forgo college—but they saw my determination and supported me in their own way.

---

**Building a Legacy**

Young adulthood was about more than just personal success. I wanted to change the trajectory for my family, to give them the security and opportunities we hadn't always had. Watching my peers' families thrive inspired me to create that kind of life for my parents and, one day, my own children.

Those years taught me the value of hard work, the power of vision, and the importance of betting on yourself. I wasn't just chasing a career—I was building a legacy, proving that you don't need a traditional path to achieve extraordinary things. And while my journey was far from conventional, it was uniquely mine—a testament to resilience, creativity, and unshakable belief in what's possible.

# CHAPTER 6
# A VISIONARY WITHOUT A DEGREE – "BREAKING THE MOLD"

**FOR A LONG TIME,** the path to success was presented as a straight line: graduate high school, go to college, get a degree, and then climb the corporate ladder. It's a respectable route, but it wasn't mine. I knew early on that I didn't fit the mold. I wasn't interested in following a formula—I was interested in results. And sometimes, the best way to get results is to take an uncharted path.

When I graduated from high school, I had my sights set on Embry-Riddle Aeronautical University. It was the dream school for anyone passionate about aviation, and I could already see myself excelling there. But reality hit hard when I looked at the costs. Tuition, books, materials, tools—it was all far beyond what I could afford. It wasn't just a financial hurdle; it felt like a wall.

I had a choice to make. I could spend years working toward the funds for a degree, or I could figure out another way to get where I wanted to go. The decision wasn't easy, but I realized something important: education doesn't always come from a classroom. Some of the greatest innovators in history—people like Garrett Morgan, Thomas

Edison, and Henry Ford—built their legacies without traditional schooling. They were proof that knowledge, not credentials, is what drives success. And so, I decided to bet on myself.

---

## The Self-Taught Path

In the early 1990s, certifications were becoming a big deal in the tech world. Companies cared less about where you studied and more about what you could do. That realization was my golden ticket. If I couldn't afford a degree, I'd earn certifications that proved I could solve problems better than anyone else.

I threw myself into learning. Nights and weekends were spent poring over manuals, disassembling computers, and troubleshooting networks. It wasn't glamorous work, but it was effective. Every time I crashed a system, I learned something new by rebuilding it from scratch. There were moments of frustration—hours spent trying to debug lines of code or figure out why a server refused to connect—but I thrived on the challenge.

One of the most transformative periods of my early career came when I pursued certifications in Microsoft, Cisco, and Linux. These weren't just pieces of paper; they were validation of my skills. Each one opened new doors. I started landing better-paying jobs, often out-earning colleagues with degrees because I could prove my worth on the spot. My employers didn't care where I learned—only that I could deliver.

---

## Breaking the Pay Ceiling

The self-taught path also taught me how to advocate for myself. I remember my first real tech job, where I was hired at an entry-level salary despite doing mid-level work. At first, I thought I just needed to work harder to get noticed. But after a year of exceeding expectations, I realized that recognition wouldn't come automatically. I needed to demand it.

Armed with my certifications and a track record of success, I scheduled a meeting with my manager. I walked in, laid out my contributions, and made a case for a raise. To my surprise, he agreed on the spot. That moment was empowering—it showed me that knowledge and confidence could break through barriers that traditional qualifications couldn't.

From that point forward, I never hesitated to ask for what I deserved. Whenever I added a new certification or solved a particularly difficult problem, I made sure my employers knew it. Over time, I developed a reputation as someone who could deliver results, and the opportunities kept coming.

---

## Crashing and Learning

Being self-taught also gave me an edge when it came to problem-solving. I wasn't confined by conventional thinking or reliant on textbooks. I learned through experimentation, which meant I understood the "why" behind every solution. That approach made me adaptable—a skill that would become invaluable in my later ventures.

One vivid memory stands out from those early days. I was working on a client's network that had completely crashed. It was a mess—half the systems wouldn't boot, and the other half were riddled with errors. The client was panicked, and I had no manual or mentor to guide me. I spent 14 straight hours troubleshooting, going through every line of code, every connection, until I found the issue: a corrupted driver that had caused a cascading failure. Fixing it was tedious, but when the system came back online, the client's relief was palpable.

Moments like that solidified my belief in my abilities. They weren't just wins—they were reminders that being self-taught didn't make me less qualified. If anything, it made me more resourceful.

---

## The Value of Unconventional Thinking

One of the biggest lessons I learned during this time was the value of thinking differently. Traditional education often teaches you to follow a set path, but the self-taught journey forces you to carve your own. That difference in mindset is what allowed me to see opportunities others overlooked.

For example, while many of my peers focused on specific niches, I made it a point to learn broadly. I studied networks, databases, hardware, and software. This interdisciplinary approach gave me a unique perspective. When problems arose, I could connect the dots in ways others couldn't, often identifying solutions that seemed obvious in hindsight.

Another key insight came from observing the tech industry itself. I noticed that companies were often reactive, addressing issues after they became problems. I decided to be proactive, predicting trends and preparing for challenges before they arose. That mindset would become a cornerstone of my entrepreneurial ventures, especially with Sikhona.

———

**Lessons for the Next Generation**

Looking back, my decision to forgo a traditional degree was one of the best choices I ever made. It wasn't the easy path, but it was the right one for me. It taught me resilience, adaptability, and the importance of believing in myself even when others doubted.

To anyone reading this, especially those who feel constrained by traditional expectations, I want you to know this: there's no one-size-fits-all formula for success. Whether you're pursuing a degree, certifications, or teaching yourself from scratch, what matters most is your willingness to learn, adapt, and push through challenges. The path may not be straightforward, but it will be yours—and that makes all the difference.

# CHAPTER 7
## VISIONARY ON THE RISE

BY MY MID-TWENTIES, my vision for what I wanted was crystal clear. Technology wasn't just my job; it was my passion. I was at Dimension Enterprises, and it was there I started to really come into my own. My job was to architect networks—big ones—for telecommunications companies, governments, you name it. They'd come to us with a problem, and it was up to my team to figure out the solution. But here's the thing: I wasn't just part of the team—I quickly became the one people looked to for answers.

I remember those early days like they were yesterday. We'd be sitting in meetings, throwing around ideas, and I wasn't afraid to challenge anyone if I thought their plan could be better. I didn't do it to ruffle feathers; I did it because I cared. And people noticed. Pretty soon, I wasn't just contributing—I was leading. It felt good to know my voice mattered, that my ideas were making a difference.

Dimension Enterprises was also where I started building a network of people who would stay with me for years. When I moved on to Verizon, I brought a couple of those guys with me. Loyalty and relationships have always been important to me. If someone shows they're willing to go the extra mile, I'll make sure to take them with me when I move forward.

When I joined Verizon, the stakes got even higher… but more on that later.

---

**Balancing Acts**

If there's one thing I've learned, it's that ambition comes at a cost. In my twenties and early thirties, I was so focused on building my career that I let a lot of other things fall by the wayside. Work came first—always. I'd spend eight hours a day at my job, then come home and put in another six to eight hours on my computer, learning, experimenting, and staying ahead of the curve. Weekends were no different. For about five years straight, that was my life.

Looking back, I can see how that single-minded focus hurt my personal relationships. I wasn't the kind of guy who went out partying every weekend or spent hours hanging out with friends. I had friends, sure, but they didn't always understand why I was so driven. Even when I was dating, I found it hard to give relationships the attention they needed. It's not something I'm proud of, but it's the truth.

Still, I wasn't all work and no play. When I did take breaks, I'd spend time with my grandparents or hit the bowling alley with my brother and friends. Bowling was one of those activities that let me unwind and disconnect from the grind. Music was another escape. My brother and I had been making music together since we were kids, and even when life got busy, we found time to record in the studio.

But even with those moments of downtime, the pressure never really went away. I felt like I was carrying the weight of not just my own ambitions but the hopes of my family too. My parents had worked hard to give me a good life, and I wanted to make sure I could give back to them and provide even more for the next generation. That drive kept me going, even when the hours were long and the sacrifices felt heavy.

The hardest part of those years was finding balance. It's something I'm still working on, honestly. But what I've learned is that you can't pour from an empty cup. If you're going to give your all to your work, your family, and your dreams, you have to take care of yourself too. It's not always easy, but it's necessary.

# CHAPTER 8
# THE FIRST BREAKTHROUGH – "FOOD ON THE MOVE"

BY THE TIME I was in my early twenties, I had a decent job and a growing skill set, but something inside me was restless. Working for others didn't give me the freedom to create, to innovate, or to solve problems the way I wanted. I needed a way to channel my energy and ideas into something bigger. That's when I stumbled into the food delivery business—by accident at first, but it quickly became my first real entrepreneurial breakthrough.

At the time, I was working as a delivery driver for a company called Take Out Taxi. The concept was simple: they partnered with local restaurants, took orders from customers, and hired drivers like me to deliver the food. It was a decent side hustle, but as I worked, I noticed inefficiencies everywhere. Orders were late because dispatchers had trouble keeping track of drivers. Communication between restaurants and drivers was clunky at best, often relying on outdated technology like pagers and handwritten notes. It was clear to me that with a few changes, the entire operation could run more smoothly.

I wasn't the type to stay quiet when I saw room for

improvement. So, I started taking my ideas to Kevin, the owner of Take Out Taxi. One day, I said, "Hey Kevin, you ever think about doing this, this, and this to help improve efficiency? We could streamline communication with the drivers, set up better tracking, and improve the process for getting orders from restaurants to customers."

Kevin would always nod politely and say, "Thanks for the ideas," but he never acted on them. It became a pattern —I'd pitch him ideas, hoping he'd see the potential, but nothing ever came of it. Then one day, after one of my usual attempts to propose improvements, Kevin said something that changed everything.

"I appreciate the ideas and all that," he said, his tone a little dismissive, "but I got this. If you want to implement your ideas, you should consider starting something like this yourself."

He didn't mean it as encouragement, but I heard it loud and clear. I nodded, smiled, and said, "You know what? You're right."

I walked out of that conversation with a newfound clarity. "I'm going to do exactly that," I thought. Right then and there, I decided to leave Take Out Taxi. I handed in my sign and uniform, thanked Kevin for the opportunity, and walked out with my mission clear: I was going to build my own food delivery service, one that ran the way I knew it should.

---

**Food on the Move Is Born**

The first thing I needed was a name. Sitting in my car later that day, brainstorming ideas, it hit me. "What am I doing here? I'm moving food!" The answer was simple and perfect: *Food on the Move*. It captured the essence of the business—fast, efficient, and straightforward. From that moment, the wheels were in motion.

Starting Food on the Move wasn't easy. I didn't have a blueprint or a lot of money, but I had a vision and a willingness to do whatever it took to make it happen. I began reaching out to local restaurants, pitching them on my idea. I explained how Food on the Move would solve the problems I'd seen firsthand—better communication with drivers, more reliable delivery times, and an overall smoother operation.

To make it work, I invested in simple but effective tools, like two-way radios to keep in touch with drivers and a basic software system to manage orders and routes. These small changes made a huge difference, and soon, Food on the Move started to gain traction. Restaurants appreciated the professionalism, customers loved the reliability, and drivers felt like they were part of a well-oiled machine.

---

### The Spark of Entrepreneurship

Looking back, that conversation with Kevin was a turning point in my life. His dismissal could have discouraged me, but instead, it lit a fire inside me. It taught me that sometimes, you don't wait for permission to pursue your vision—you give yourself the green light.

Food on the Move became more than just a business; it became proof that I could take an idea, build it from the ground up, and turn it into something successful. It was my first real taste of entrepreneurship, and it showed me that I didn't need to rely on someone else's approval to create something meaningful.

Kevin's words, even though he didn't intend them to be, were the best advice I could have gotten: *If you think you can do better, start your own company.* And so, I did. That decision set the stage for everything that came after, teaching me lessons that would shape the rest of my journey.

---

### The Birth of Food on the Move

Launching a business is never easy, especially when you're young and resources are limited. I didn't have investors or a safety net—I had determination and a willingness to work hard. The first step was figuring out how to stand out from the competition. Take Out Taxi relied heavily on two-way radios to communicate between dispatch, drivers, and restaurants. It was a cumbersome system that created delays and left plenty of room for error. I saw an opportunity to do things differently.

Instead of radios, I decided to use cell phones. It sounds like a no-brainer today, but in the early '90s, cell phones were a luxury, and airtime wasn't cheap—it cost over a dollar per minute. Still, I believed the investment was worth it. Direct communication between drivers and restaurants would streamline operations and improve customer satisfaction, giving Food on the Move a competitive edge.

The next challenge was building relationships with local restaurants. I started small, reaching out to owners and managers in my community. My pitch was simple: "Let me handle your deliveries so you can focus on what you do best—making great food." It took a lot of persistence, but eventually, I secured a handful of partnerships. With a small team of drivers and a growing list of clients, Food on the Move was officially in business.

---

**Growing Pains and Early Wins**

The early days were a mix of excitement and chaos. I wore every hat imaginable—dispatcher, driver, customer service rep, and marketer. I worked 16-hour days, juggling orders, managing drivers, and fielding complaints. There were times when I felt completely overwhelmed, but the thrill of building something from scratch kept me going.

One of the biggest lessons I learned during this time was the importance of adaptability. No matter how carefully I planned, something always went wrong. Drivers got lost, restaurants ran behind on orders, and customers were quick to voice their frustrations. But each setback was an opportunity to refine our processes. I started implementing systems to track orders more efficiently and trained my drivers to handle common issues on the spot. Slowly but surely, we began to gain a reputation for reliability.

One of our earliest wins came from a partnership with a popular local Italian restaurant. They had been working with Take Out Taxi but were frustrated with late deliveries and poor communication.

I promised them that Food on the Move would do better—and we did. Within weeks, they saw a noticeable uptick in customer satisfaction, and word began to spread. Other restaurants took notice, and before long, we were expanding into new neighborhoods. *We were on the move!*

---

## Scaling Up

As Food on the Move grew, so did the challenges. We expanded into Fairfax, Loudoun, and Prince William counties, covering a large portion of Northern Virginia. At our peak, we had over 130 drivers on the road, handling thousands of orders each day. Managing such rapid growth was exhilarating, but it also tested my leadership skills in ways I hadn't anticipated.

One of the biggest hurdles was logistics. Coordinating dozens of drivers across multiple counties required more than just cell phones—it required strategy. I began experimenting with scheduling software, trying to optimize routes and minimize wait times. I even considered hiring a software developer to build a custom dispatch system, but the costs were prohibitive. Instead, I relied on a combination of intuition and trial and error to keep things running smoothly.

Hiring and retaining good drivers was another challenge. The job wasn't easy—long hours, tight deadlines, and demanding customers made it tough to find people who were both reliable and motivated. To address this, I created an incentive program, rewarding top performers with bonuses and flexible schedules. It wasn't a perfect solution, but it helped build a team that was committed to the company's success.

---

## A Taste of Racism

As Food on the Move grew, so did my experiences with the harsh realities of racism—personal and professional moments that left an indelible mark. Success didn't shield me from prejudice; if anything, it seemed to magnify it.

These encounters weren't just isolated events—they were reminders of the systemic challenges I would face, no matter how much I accomplished.

One of my earliest memories of racism still haunts me. I was in the fourth grade, walking back to school to retrieve a book I had forgotten. A pickup truck with two white men slowed down as they passed. One of them leaned out the window and hurled a piece of fruit at me, hitting me squarely in the stomach. The pain wasn't just physical—it was an introduction to a world where my skin color made me a target.

I never told my parents about it. For years, I buried that memory, only speaking about it decades later. "At the time, I didn't feel that it did a whole lot," I said. "It seemed like it subsided and kind of went to the back of my mind. But as I got older and had a chance to experience racism more, I saw that we're still in the fight."

As I built Food on the Move, I began to see how those early experiences of racism connected to the challenges I faced as a Black entrepreneur. One encounter stands out—a moment that underscored how deeply entrenched racism can be, even when I was at the height of success.

---

### Pulled Over, Doubted, and Dismissed

One day, as I was making my rounds, transporting cash and checks from Food on the Move to the bank, I was pulled over by the police. At first, I thought it was a routine stop, but it quickly became clear that this was something else entirely.

The officer glanced into the backseat, where the money

and checks were stacked, ready for deposit. His eyes narrowed. "What's all this?" he asked.

"It's the day's earnings," I replied. "I'm taking them to the bank."

He scoffed. "From what business?"

I pointed to the logo on the envelopes. "Food on the Move."

The officer's expression hardened. "That's not your business," he said flatly. "We order from them all the time. You don't own Food on the Move."

I stared at him, incredulous. "I do own it. I built it."

But he wasn't having it. "The only thing you're going to own is a cell," he sneered.

They put me in handcuffs, ignoring my protests and the evidence I provided. The officer turned to his partner. "We'll call the owner and let them know we found their money and checks," he said, with a smirk that made it clear he didn't believe a word I'd said.

"I am the owner," I repeated, my voice steady even as anger simmered beneath the surface. "You're not going to get anyone else on the phone because I'm standing right here."

They still took me in. I sat in a holding cell for hours, treated like a criminal despite having done nothing wrong. Eventually, they released me, but not before confiscating the money and checks. "We'll follow up with Food on the Move tomorrow," one officer said as he handed me my release papers. "We'll come looking for you if anything doesn't check out."

The next day, I returned to the station armed with a letter from my bank confirming the deposits and my ownership of the business. The magistrate and sergeant returned my money and checks without an apology. "Here you go," they said, as if they had done me a favor. It was a bitter pill to swallow.

---

### **The Weight of Success**

That wasn't the only time I was targeted. There were countless other moments—being pulled over in a nice car and asked, "How can you afford this?" or questioned about why I lived in a predominantly white neighborhood. The insinuation was always the same: success didn't belong to me.

"Even as a businessman, very successful and everything else," I explained, "the police would tell me things like, 'You people never have ID,' or find any reason to arrest me. It showed me that no matter what I accomplished, I would still be seen as less than in their eyes."

But even in the face of such blatant racism, I refused to let it change how I interacted with the world. "I still got along with white people," I said. "Because I knew some good white people. But it allowed me to sit back and analyze to see who they really were. Even the ones who smiled in my face often carried this underlying tone of superiority, like they were better or had more."

Not every relationship was like that, of course. "Some white boys I knew were so cool, they didn't even know they were white," I joked. But I never let my guard down completely. Experience had taught me to recognize the difference between genuine friendship and a "pet project"—a Black friend some people kept around to prove they weren't racist.

---

## Resilience Through It All

Those encounters didn't break me—they fueled me. Each experience was a reminder of what I was up against and why I needed to keep pushing forward. Racism wasn't going to stop me from achieving my goals, nor was it going to define how I lived my life.

"I became sensitive to it over the years," I said, "but I didn't let it change how I acted toward them. I always heard, 'Things are better, things are fair,' but I knew the truth. And that truth made me even more determined to succeed."

Building Food on the Move wasn't just about creating a business—it was about proving that success belonged to me, no matter what anyone said or thought. Those moments of racism, painful as they were, became stepping stones on the path to something greater. They reminded me of the resilience I carried within me and the importance of using that strength to fuel my future.

---

## The Sale

Despite our success, running Food on the Move eventually took a toll on me. The long hours, constant problem-solving, and sheer scale of the operation left little room for anything else in my life. I loved what I had built, but I also knew it wasn't sustainable for the long term—not without significant investment and infrastructure upgrades.

That's when I decided to sell. A competitor approached me with an offer to buy Food on the Move, and after some negotiation, we reached a deal. Letting go of the company was bittersweet. On one hand, I was proud of what I had accomplished. On the other, I couldn't shake the feeling that I was leaving something unfinished. But I knew it was the right decision for me at the time.

---

**Lessons Learned**

Food on the Move was more than a business—it was a crash course in entrepreneurship. It taught me the value of persistence, the importance of listening to customers, and the power of innovation. Most importantly, it showed me that I had what it took to compete with anyone, regardless of resources or experience.

Looking back, I realize that Food on the Move was the foundation for everything that came after. It gave me the confidence to tackle bigger challenges and the skills to navigate the unpredictable world of business. And while it was only the beginning, it was a beginning that shaped the rest of my journey.

# CHAPTER 9
# BUILDING A PROTOTYPE THAT CHANGED IT ALL – "STREAMNOW"

BY THE LATE 1990S, the world was on the brink of a digital revolution. The internet was no longer just a tool for sending emails or browsing static web pages—it was becoming a platform for rich multimedia experiences. I could see where things were heading: faster connections, better compression algorithms, and a growing hunger for video content. It wasn't just a trend; it was a tidal wave, and I wanted to be at the forefront.

At the time, streaming video was more science fiction than reality. The technology existed, but it was clunky, unreliable, and nowhere near ready for prime time. Companies like Bell Atlantic were pouring millions into solving the problem, but they kept running into the same roadblocks: buffering, poor resolution, and exorbitant costs.

Meanwhile, I was working quietly in my own corner, convinced that I could crack the code.

### The Birth of an Idea

It was a brisk afternoon in 1999, and the air carried the

faint aroma of cigarettes and fresh coffee as workers milled about during their break at Qwest LCI. I stepped outside for a breather when I spotted my friend, Seong Wong, leaning against a brick wall, his cigarette lazily trailing smoke into the sky. Seong was a sharp guy, he worked in the sales department. But today, he looked irritated, puffing away with the kind of intensity that signaled something was bothering him.

"Seong, what's up?" I asked, joining him.

He sighed, flicking the ash from his cigarette. "Man, my internet is driving me crazy. It keeps cutting out, dropping connections in the middle of downloads, and don't even get me started on trying to stream anything. It's a nightmare."

Streaming. That word stuck in my head like a song on repeat. Streaming video was still a pipe dream for most people—a futuristic concept that seemed as far off as flying cars.

But as Seong ranted about the frustrations of unstable connections, an idea began forming in my mind. What if I could stabilize the internet? What if I could create a solution that didn't just fix the problem but opened the door to something bigger?

I leaned against the wall next to him, the wheels in my head already turning. "What if you didn't have to deal with that? What if the internet could handle video—movies, shows, everything—without stuttering or crashing?"

Seong laughed, shaking his head. "Yeah, sure. When pigs fly."

But I wasn't joking. "I'm serious. I think I can do it."

## Cracking the Code

That conversation with Seong lit a fire inside me. I

couldn't stop thinking about it—the idea of not just fixing the internet but transforming it. I didn't have a team or a roadmap, but I had determination. And sometimes, that's all you need to get started.

I spent the next 30 days in my apartment, pouring over technical papers, sketching diagrams on napkins, and running tests on a second-hand computer I'd cobbled together. I didn't just want to stabilize internet connections; I wanted to build a platform that could deliver high-quality video seamlessly. If I could crack this code, it wouldn't just solve Seong's problem—it would change the game.

Every night, I worked late into the evening, my desk cluttered with empty coffee cups and scraps of paper. The glow of the monitor lit up the room as I wrote lines of code, tweaking and testing until my eyes burned from staring at the screen. But no matter how tired I was, I couldn't stop. I was chasing something bigger than myself—a vision of what the internet could be.

---

## The Breakthrough

I'll never forget the night it finally clicked. It was past midnight, and the world outside was silent except for the hum of the occasional car. I had just run another test—a small video clip streamed over my makeshift setup. As I watched the video play smoothly, without a single glitch or stutter, a chill ran down my spine.

"I did it," I whispered, leaning back in my chair, a mixture of disbelief and exhilaration washing over me. "I actually did it."

The solution wasn't just a patch for unstable connections; it was a whole new way of thinking about data delivery. By combining innovative compression techniques with dynamic caching, I had created a system that could adjust to changing network conditions in real-time. It was fast, efficient, and, most importantly, it worked.

I sat there for a moment, staring at the screen, letting it sink in. This wasn't just a prototype—it was a glimpse into the future. A future where people could stream movies and shows from anywhere, without worrying about buffering or crashes. A future where the internet wasn't just a tool for communication but a platform for entertainment, education, and connection.

## The Next Step

The next morning, I called Seong. "Remember when you said pigs would fly before the internet could handle streaming?"

He laughed. "Yeah, why?"

"Well," I said, grinning, "you might want to look outside. Because I think I just built the wings."

That breakthrough was the start of everything. It wasn't easy—there were countless challenges and setbacks along the way—but that moment proved to me that innovation is possible when you refuse to settle for less. It also taught me the power of listening—to frustrations, to needs, and to the possibilities they reveal.

And to think, it all started with a cigarette break and a conversation. Sometimes, the most revolutionary ideas come from the most ordinary moments.

## The Vision

My idea for StreamNow wasn't just about technology; it was about accessibility. I envisioned a world where anyone, anywhere, could stream high-quality video with minimal hassle. Movies, TV shows, educational content—it was all within reach if the infrastructure could catch up to the demand. What the industry needed was a solution that was scalable, cost-effective, and easy to implement. I believed I could create it.

The first step was understanding the problem at its core. Streaming wasn't just about delivering data; it was about doing it in a way that minimized latency, maximized quality, and kept costs manageable. That meant tackling issues like compression algorithms, packet loss, and bandwidth optimization—all while ensuring the system could handle millions of simultaneous users.

I didn't have a team of engineers or a research lab at my disposal. What I did have was determination and a willingness to experiment. I spent countless nights reading technical papers, studying existing solutions, and sketching out ideas on scratch paper. Each failure brought me closer to a breakthrough, and I could feel the pieces of the puzzle starting to come together.

---

## The Prototype

Within 30 days, I had a working prototype. It wasn't perfect, but it was functional—and, more importantly, it was proof that my ideas could work in the real world. The prototype used a combination of proprietary compression techniques and intelligent caching to deliver video streams with minimal buffering. It wasn't just faster; it was smarter, dynamically adjusting to changes in network conditions to ensure a seamless experience.

I still remember the first time I tested it. Watching a video stream smoothly, without the stuttering and pixelation that plagued other solutions, was like seeing the future unfold in real-time. It wasn't flashy, but it worked. And in a world where even the biggest players were struggling to make streaming viable, that was enough to get people's attention.

---

**The Bell Atlantic Acquisition**

As word of the prototype spread, it caught the attention of OnePoint Communications, a telecom company looking for an edge in the competitive internet landscape. They saw potential in StreamNow and acquired the technology as part of a broader strategy to pitch themselves as a leader in next-generation connectivity.

Not long after, Bell Atlantic acquired OnePoint Communications. At the time, Bell Atlantic was positioning itself to become a major player in the internet revolution. They were willing to invest in innovative solutions, and StreamNow fit perfectly into their vision. As part of the deal, I became Vice President of Verizon Video Services, a role that would give me the platform to bring my vision for streaming to life on a much larger scale. I was thirty-one years old and eager to bring this vision to reality.

# CHAPTER 10
# THE BIG PLAYERS TAKE NOTICE – "WINNING WHERE GIANTS FAILED"

IN THE LATE '90S, Bell Atlantic Video Services (BVS) had already sunk millions into the Stargazer Project, an ambitious but faltering attempt to bring video-on-demand to the masses. The company had spent over $80 million trying to perfect the technology, but by the time I stepped into the picture, their efforts had produced little more than frustration and empty promises.

It wasn't long before I realized that what they couldn't achieve with all that money and manpower, I could do with the right vision and determination. This wasn't arrogance—it was clarity. The technology was evolving, and I understood it on a fundamental level in a way their team, mired in legacy systems and bureaucracy, simply didn't.

---

### The Visit to BVS

I was sitting in my office in Herndon, Virginia, when it struck me that I needed to see for myself what BVS had been working on. I called my admin and said, "Find out who's running things at Bell Atlantic Video Services." She came back quickly with the names, and I turned to one of my directors. "Set up a meeting for us this afternoon. Tell them this is coming directly from the Vice President of Video Services."

It was a power move, and I knew it. When they tried to push the meeting out by days, my director repeated the line: "This is coming directly from the Vice President. If you have an issue with it, you can take it up with him." That settled it. They agreed to meet.

I took three of my directors and one of my engineering guys over to BVS. The tension in the air was palpable from the moment we walked in. I could feel the undertones—the unspoken questions hanging in the room: *Who is this young Black man walking in here, calling the shots?*

We asked for a tour, and they begrudgingly obliged. As we walked through their operations, they described the Stargazer Project, detailing how they'd been working on it since the late '80s. "We've spent about $80 million on this over the years," one of their managers said, as if that was supposed to impress me.

"And what have you accomplished?" I asked.

They explained that they'd been trying to secure content deals with the major movie studios but had run into roadblocks at every turn. "They're very picky," one of them admitted, a hint of defeat in his voice. "They're not willing to take risks."

I nodded, keeping my thoughts to myself. They'd poured millions into this project and still didn't have a workable solution or content to show for it. As we wrapped up the tour, I turned to my team. "We're not coming back here," I told them later. "I have a plan."

## Taking Matters into My Own Hands

I wasn't going to let politics or hesitation hold me back. If BVS couldn't get the studios on board, I would. The technology I had developed was ahead of its time—efficient, scalable, and ready to deliver high-quality video content across regular copper lines. The studios needed to see it in action.

I gathered my team and laid out the plan. "We're going to Hollywood," I said. "We're going to meet with Sony, Disney, DreamWorks, New Line Cinema, Warner Brothers, MGM—every major studio we can get in front of."

To make this happen, I tapped Dennis Sullivan, a seasoned industry professional with connections to the studios. He flew into D.C., and together, we strategized. "Can you get us an audience?" I asked him. Dennis, confident in my reputation and the Verizon name, assured me he could.

We decided to host the demonstration at the Regent Beverly Wilshire Hotel in Beverly Hills—a venue as iconic as the names on our guest list. I told my admin to book a conference room that could hold about 30 people, along with six rooms for my team for the week. The setup had to be flawless. We weren't just selling technology; we were selling trust and a vision for the future.

## The Hollywood Pitch

The first day of the demonstration arrived, and nearly all the studios showed up—everyone except Disney, who initially wanted no part of what we were presenting. Still, I wasn't discouraged. I knew the quality of our system would speak for itself. Before starting the formal presentation, I decided to build some suspense. My team and I mingled quietly while the studio representatives helped themselves to the refreshments.

A few minutes before the official start time, people began murmuring, "Where's Michael Thompson? Is he late?" I gave my team a knowing look and waited a beat before standing up. "Welcome," I said. "I'm Michael Thompson, Vice President of Video Services at Verizon." The room fell silent as they registered who I was. Their surprise was obvious—I didn't fit the mold of what they were expecting.

As we kicked off the demonstration, I pointed to the video that had been playing on a monitor since the guests arrived. "This video you've been watching," I said, "is streaming directly from our system over these two copper wires." One skeptical executive leaned forward. "You're saying if I cut this wire, the video will stop?"

"Absolutely," I replied. "Go ahead."

He pulled out a pair of scissors and cut the wire. The video froze instantly. "Now," I said, "twist the wires back together and refresh the browser." He followed my instructions, and the video resumed seamlessly. The room erupted in murmurs of disbelief and admiration.

---

**Turning the Tide**

The next day, the momentum shifted. Representatives from Disney, who initially declined to attend, showed up with their technical teams. Word had spread about the demonstration, and suddenly, everyone wanted to see what we had built. Over the next two days, we fielded relentless questions from studio engineers and executives, addressing every concern about security, scalability, and billing.

By the end of the third day, the studios were ready to sign deals. "How do we make this happen?" they asked. My answer was simple: "We'll put money in escrow so you can draw down whenever users access your content. For the first year, you keep all the profits. All we need is the cost of delivery."

The studios were thrilled. They saw the potential, the vision, and the technology that could finally bring video-on-demand to life. But back at Verizon, not everyone shared their enthusiasm.

---

**The Fallout**

When I presented the deal to Verizon's leadership, my proposal was met with resistance. "We need to make money on this," my boss, Mark Fuller, said. "This is too risky." I argued that the priority should be getting the studios on board and proving the concept. "If we don't move now," I warned, "someone else will."

Ultimately, Verizon vetoed the deal, prioritizing short-term profits over long-term vision. Disillusioned, I knew it was time to move on. My team scattered—some left the company, others were reassigned—and the opportunity slipped away.

Years later, as companies like Netflix and Apple dominated the streaming industry, I couldn't help but think back to those days at BVS and the missed chance to lead a revolution. But for me, the experience was invaluable. It proved that with vision, determination, and the right team, even the giants can be challenged—and sometimes, they can fall.

---

## What Could Have Been

My prototype was up and ready to be used in the mid-2000s, nearly seven years before Netflix began its foray into streaming in 2007. It wasn't just ahead of its time; it was perfectly timed to capitalize on an industry that was just beginning to realize the power of online video. By leveraging the relationships I had cultivated with the major studios, Verizon could have established itself as the pioneer in this space.

From there, my next plan was to expand into music, securing deals that could have positioned Verizon as a competitor in a space Apple now dominates. Instead, companies like Netflix, which as of this writing has a market valuation of $299 billion, stepped into the void. And Apple, sitting at a staggering $3.44 trillion market cap, took the lead not just in technology but in cultural dominance.

It's not just about the money—it's about what Verizon could have been. They had the infrastructure, the brand, and the relationships to own the streaming space long before anyone else. Instead, they let the opportunity slip away, distracted by ventures that didn't have the same long-term potential.

---

## Lessons from the Missed Opportunity

Looking back, the most frustrating part wasn't the failure itself—it was knowing how close we came to success.

My prototype and the pitch to the studios proved that streaming wasn't just a dream; it was achievable. The technology was ready, the studios were on board, and the audience was hungry for what we had to offer.

The lesson here is simple: vision without action is just a dream. It's not enough to see the potential of an idea—you have to commit to it fully. Another critical lesson I learned was the importance of maintaining control over my ideas and creative direction. After selling the rights to my prototype and stepping into the role of Vice President at Verizon, I lost control over how my vision would unfold. Whatever political or organizational dynamics caused the deal to stall, I realized that my hands were tied. I could pitch ideas, even shape strategy, but the final decisions were no longer mine to make.

That realization left a mark. I vowed that things would be different the next time. This is one of the reasons why I've taken a firm stance in how I manage my businesses today. It's also why I turned down a $200 million offer from a highly respected venture capital firm to raise money for my current company, Sikhona. As tempting as the offer was, I knew that accepting it would mean ceding control of my vision. I've also turned down offers to acquire Sikhona for the same reason. I want to see my baby grow to full maturity, guided by my direction and not surrendered to someone else's priorities.

---

## Moving Forward

StreamNow wasn't just a prototype—it was a glimpse into a future that I still believe is achievable. And while Verizon chose a different path, the lessons I learned during that time have stayed with me. They shaped my approach to innovation, my belief in the power of vision, and my determination to never let another opportunity slip away. Most importantly, they reinforced my commitment to always steer my own ship, ensuring that the ideas I create remain true to their purpose and potential.

Professionally, the missed opportunity at Verizon was a turning point for me. It was the first time I truly felt the weight of what it meant to be part of a massive corporation. On paper, I had a significant title and influence, but in reality, my ability to execute my vision was limited by layers of bureaucracy. It was a hard pill to swallow, but it taught me an invaluable lesson: control is everything. No matter how brilliant your ideas are, if you don't have the power to implement them, they may never see the light of day.

Personally, the experience left me more determined than ever to chart my own course. I realized that if I wanted to bring my ideas to life, I needed to do it on my own terms. That realization was one of the driving forces behind the founding of Sikhona, where I made a promise to myself that I would never relinquish control of my vision again.

---

## The Resilient Realist

After Verizon vetoed the deal and the opportunity with the studios slipped away, many might have been devastated. For me, though, the situation was different. I've always been a realist, someone who looks at the facts, acknowledges them, and moves forward. I wasn't about to let one setback define me or derail my future.

"This is not my company," I told myself. "They can do with it as they see fit." It was a simple yet powerful truth that helped me maintain perspective. If I wanted the final say in those decisions, I realized, I should have never sold my prototype in the first place. That's a hard lesson, but one I took to heart. I processed it, accepted it, and moved on.

"After a day or two of thinking about this thing, it was a done deal," I explained. There was no drawn-out grief or wallowing in frustration. I knew my value, and I knew my ability to create and innovate hadn't been diminished. The decision to walk away wasn't easy, but it was necessary. I wasn't going to stay in an environment where my vision was stifled.

What kept me grounded was the confidence that I could come back stronger. "There was no damage done that I could not recover from," I said. That resilience wasn't just a coping mechanism—it was a driving force. It reminded me that setbacks are temporary, and with the right mindset, they can become opportunities for growth.

# CHAPTER 11
# ENTREPRENEURIAL VENTURES – "BEYOND THE PROTOTYPE"

AFTER MY TIME AT VERIZON, I found myself at a crossroads. The missed opportunity with StreamNow left a bitter taste, but it also lit a fire inside me. I knew I had the skills and vision to create something meaningful—I just needed the right platform. The corporate world, with all its politics and bureaucracy, wasn't the place for me. If I wanted to shape the future, I would have to do it on my own terms.

That decision marked the beginning of my journey as an independent entrepreneur. Over the next two decades, I would launch and lead a series of ventures, each one teaching me invaluable lessons about innovation, leadership, and resilience. Some were wildly successful, while others fell short of expectations. But every step prepared me for the vision I would one day bring to life with Sikhona.

---

**The First Leap: Food on the Move**

Long before StreamNow, I had my first taste of entrepreneurship with Food on the Move, a delivery service I started in the early '90s. It was essentially a precursor to modern apps like DoorDash or Uber Eats, and it gave me a crash course in building and scaling a business from scratch. I had no investors, no roadmap, and no guarantees—just a vision and the willingness to work harder than anyone else.

Food on the Move taught me the importance of adaptability. Every day brought new challenges, from managing drivers to handling customer complaints. But it also taught me the power of persistence. By the time I sold the business, we had expanded into multiple counties and built a reputation for reliability. That early success gave me the confidence to take bigger risks in the future.

---

### The Innovator's Mindset: StreamNow

StreamNow was the next big leap. While the prototype didn't reach its full potential at Verizon, it proved that I could think ahead of the curve and execute on bold ideas. It also showed me the value of scalability—creating solutions that don't just solve problems for today, but anticipate the needs of tomorrow.

More importantly, StreamNow reinforced my belief in the power of innovation. Even though the corporate structure at Verizon stifled the project, the technology itself was sound. That validation fueled my drive to keep creating, even when the odds seemed stacked against me.

---

### Create A Disc: The Digital Revolution

In the late '90s, I turned my attention to digital storage.

Create A Disc was a platform designed to help users burn their own CDs and DVDs, a technology that was just starting to take off at the time. It wasn't just about convenience—it was about empowerment. For the first time, people could create and share their own digital content, from family photos to home videos.

The project gained significant traction, and at one point, we were close to closing a major deal with Kodak. While that deal ultimately fell through, the experience taught me the importance of negotiation and partnership. It also showed me that even when things don't go as planned, the effort and connections you build can lead to future opportunities.

---

### Radio DVR: Capturing the Moment

In 2007, I launched Radio DVR, a subscription service that allowed users to record and save radio broadcasts for later listening. At its peak, the platform had over 150,000 monthly subscribers, and it was a testament to my ability to identify and capitalize on emerging trends. Radio DVR wasn't just about technology—it was about giving people control over their media consumption.

This venture also taught me the importance of listening to your audience. Many of our features were developed in response to user feedback, and that responsiveness helped us stand out in a crowded market. While I eventually moved on from the project, the experience reinforced my belief that customer-centric innovation is key to long-term success.

## InPromixity: Connecting Communities

By the early 2010s, I was exploring ways to use technology to bring people together. InPromixity was a platform designed to connect users with local businesses and services, creating a digital ecosystem that fostered community engagement. InProximity transformed the guest experience by providing hotel visitors with an innovative solution to discover nearby restaurants and points of interest more accurately and efficiently than traditional GPS or Google Maps. By tapping an NFC-enabled phone on strategically placed posters within hotel rooms or at the front desk, guests gained instant access to curated recommendations—elevating convenience to an entirely new

I licensed it to a leading global hospitality company under an exclusive 10-year non-compete agreement, securing a lucrative deal that positioned InProximity as the go-to solution for hotel-based point-of-interest navigation.

InPromixity was one of my most ambitious projects, and it taught me valuable lessons about scale. Coordinating between investors, developers, and end-users required a level of precision and communication that I hadn't experienced before. It also reminded me that even the best ideas need a solid team to bring them to life. While InPromixity eventually wound down, the lessons I learned from that experience were instrumental in shaping my approach to leadership.

---

## OBT Social: The Path to Sikhona

In many ways, OBT Social was the precursor to Sikhona. Short for "Our Black Truth," the platform was designed to provide a safe space for Black voices to connect, share, and grow without fear of censorship. It was a bold idea, but one that faced significant challenges. Google removed the app multiple times, and we encountered resistance even within our own community.

Rather than let those setbacks derail us, I saw them as an opportunity to refine the vision. OBT Social evolved into Sikhona, a platform with a broader mission to empower marginalized communities worldwide. The name itself, a South African term meaning "We are here to be seen," reflects the resilience and determination that have defined my journey.

**Lessons from the Journey**

Each of these ventures taught me something unique. From Food on the Move, I learned the importance of adaptability and hard work. StreamNow reinforced my belief in the power of innovation, while Create A Disc and Radio DVR showed me the value of listening to your audience. InPromixity taught me how to scale effectively, and OBT Social reminded me of the importance of staying true to your mission, even in the face of adversity.

But the most important lesson was this: success is rarely a straight line. There are twists, turns, and setbacks along the way, but each one brings you closer to your goal. The key is to keep moving forward, to learn from every experience, and to never lose sight of your vision.

**Building the Foundation for Sikhona**

By the time I launched Sikhona, I felt like everything I

had done before was leading up to this moment. Each venture, with its successes and failures, had given me the tools I needed to build something truly transformative. Sikhona isn't just a company—it's a culmination of decades of learning, experimenting, and growing.

Looking back, I'm grateful for every step of the journey. The challenges I faced made me stronger, the risks I took made me wiser, and the lessons I learned made me the leader I am today. And while the road ahead will undoubtedly bring new challenges, I'm more confident than ever that Sikhona is ready to change the world.

# CHAPTER 12
# THE BIRTH OF SIKHONA – "MORE THAN JUST A COMPANY"

SIKHONA WASN'T JUST the next step in my entrepreneurial journey—it was the culmination of everything I had learned, experienced, and dreamed of over the years. By the time I launched the platform, I wasn't just building a business; I was building a movement. Sikhona was designed to be a testament to resilience, innovation, and the unyielding spirit of those who refuse to be ignored.

The name itself holds profound significance. Sikhona is a South African term that translates to "We are here to be seen." For me, it was more than a name—it was a declaration. It was a message to the world that no matter how many times history has tried to erase or marginalize certain voices, we are still here. We are thriving, innovating, and claiming our rightful place at the forefront of technology and culture.

Sikhona is a dynamic media and communications company that goes beyond boundaries. From developing cutting-edge software applications to building robust internet infrastructures, we are the architects of tomorrow's digital landscape. *We're here for the world to see!* Our content, both informative and entertaining, resonates with audiences across the globe. We're not just a company; we're a movement, reshaping how the world engages with technology and each other.

## A Vision for Africa's Digital Future

Now is the time for a company and idea like Sikhona. The world is witnessing unprecedented technological advancements, and Africa is poised to play a pivotal role in this digital transformation. With the youngest population globally and a rapidly growing workforce, Africa is brimming with potential. Yet, the digital divide continues to hinder many from accessing opportunities that technology can unlock.

Sikhona emerged as a response to this urgent need. It's not just about creating another tech platform—it's about empowering communities with the tools and resources they need to thrive in the digital age. From education and entrepreneurship to creative expression, technology can be the bridge to a brighter future. For Africa, this means leveraging its vast talent and resourcefulness to shape its own narrative on the global stage.

The vision for Sikhona goes beyond traditional boundaries. It's about harnessing technology to connect and uplift, enabling communities to solve their own challenges and create solutions that resonate locally and globally. This is why Sikhona's mission is so timely and critical. By investing in digital ecosystems that prioritize inclusivity and innovation, Sikhona aims to position Africa as a leader in the next wave of technological progress.

## A Digital Ecosystem

Sikhona is more than just a single platform; it's an interconnected ecosystem of ten distinct platforms, each designed to solve real-world problems and create opportunities. There's **Heyyo**, our answer to YouTube, which empowers creators to share their work and monetize their content. There's **DoYou**, a TikTok comparable app that offers creators fair compensation and a safer, more inclusive environment.

We also have **File Mule**, a service that allows users to send and store large files securely. **Sikhona University** provides e-learning resources to help users upskill and advance their careers. And these are just a few of the platforms we've developed, each one designed to empower individuals and communities in meaningful ways.

What sets Sikhona apart is its holistic approach. Each platform is a piece of a larger puzzle, working together to create a comprehensive digital infrastructure. It's not just about providing tools—it's about creating a support system that enables people to thrive in every aspect of their lives.

---

## Building a Movement

From the beginning, I knew Sikhona had to be more than just a company. It had to be a movement. I didn't want it to be another tech startup chasing trends—I wanted it to be a force for change. Sikhona is about rewriting the narrative, about showing the world that innovation and leadership come from all corners of the globe, including those often overlooked or underestimated.

One of our core missions is to connect the African diaspora with the continent itself. Africa is home to some of the world's fastest-growing economies and youngest populations, yet it's often excluded from global conversations about technology and innovation. Sikhona aims to change that by creating platforms that amplify African voices, showcase African talent, and build bridges between Africa and the rest of the world.

---

**Investing In Africa's Future**

Investing in Africa is something that I hold close to my heart. One of the most common complaints I hear from Africans is that they often hear promises from the African Diaspora about bringing change and opportunities to the continent, but those promises never seem to be fulfilled. That frustration is real, and it's one of the key reasons I'm proud to position Sikhona differently. We're not just talking about the future—we're already making it happen.

At Sikhona, we've taken tangible steps to show our commitment to Africa. We've already hired a growing number of talented individuals across the continent, paying them weekly and ensuring that they're compensated fairly, something that is often overlooked in many industries. We're not waiting for some distant day when we hope to bring jobs to the continent. We're doing it right now. We believe in the power and potential of the motherland, and we're making sure that belief translates into action.

Africa is the cradle of all of civilization, and it's a place of immense potential. We know that the world has long recognized this potential, but unfortunately, that recognition has often come with exploitation rather than empowerment. Now, we're here not just to be a part of Africa's future, but to be seen as an active player in shaping it. We're not here to take from Africa; we're here to give back, to uplift, and to contribute to the continent's growth in a meaningful and sustainable way.

Sikhona is more than just a company—it's a movement. As part of this movement, we're focused on creating platforms that not only benefit those within our organization but also the wider African community. One of the key ways we're doing this is through our Sikhona Summits, which are designed to bring together people from the African Diaspora and across Africa itself. These summits will focus on crucial areas like technology, entrepreneurship, innovation, and education—fields that are critical to Africa's future. By hosting these events, we aim to create a space where knowledge, resources, and opportunities can be shared across borders, empowering individuals and businesses alike.

The Sikhona Summits aren't just about networking or speeches—they're about sparking real, lasting change. They will provide a platform for Africans from all corners of the world to come together, exchange ideas, and create collaborations that push the boundaries of what is possible. We're committed to strengthening this movement and ensuring that the next generation of African leaders, entrepreneurs, and innovators have the resources and support they need to succeed.

This is just the beginning. We know the world is watching Africa's rise, and we are proud to be a part of that. With Sikhona, we're not just making promises—we're taking action, and we're excited to see where this journey will take us and the continent. Together, we'll ensure that Africa's future is as bright as its past.

---

## Facing Resistance

Like any ambitious venture, Sikhona has faced its share of challenges. Some people doubted the vision, questioning whether a company with such a broad mission could succeed. Others tried to dismiss Sikhona as "too idealistic" or "too niche." But I've never been one to back down from a challenge. Every piece of resistance we've encountered has only made me more determined to prove the doubters wrong.

Even funding Sikhona has been a journey. I've turned down offers from venture capital firms and potential acquirers because I believe in maintaining control over the vision. At one point, I rejected a $200 million investment because it came with strings that would have compromised Sikhona's mission. I've had opportunities to sell the company outright, but I know that letting go would mean sacrificing the integrity of what we've built.

Sikhona isn't just a company to me—it's a legacy. It's a reflection of everything I've fought for and everything I believe in. And I'm not willing to surrender that to anyone.

---

## The People Behind Sikhona

Sikhona wouldn't be what it is today without the incredible team behind it. From engineers and marketers to community managers, developers, and investors, every single person who contributes to Sikhona shares the same unwavering passion for creating a better future. We're more than just colleagues—we're a family, bound together by a shared vision of empowerment, progress, and impact. Each individual brings their unique strengths and perspectives, and together, we create something that goes beyond technology. We're building a legacy.

One of the most rewarding aspects of leading Sikhona has been watching our team grow, evolve, and thrive. It's been inspiring to see how we've cultivated a culture that values innovation, collaboration, and mutual respect. At Sikhona, everyone's voice is heard, and everyone's ideas are not only welcomed—they're actively encouraged. This environment has allowed creativity and innovation to flourish in ways that wouldn't have been possible without such a strong, diverse group of people working together.

But it's not just about building platforms or launching new initiatives; it's about investing in people and creating a space where individuals can truly thrive. I'm proud to say that some of the people who are with us now have been part of this journey from the very beginning—many of them, in fact, have been with me since my high school days. Their loyalty and belief in the mission have been instrumental in shaping the company's culture and success. And then there are the new faces who have recently joined us, bringing fresh perspectives and new energy to the team. They've seamlessly become part of the fabric of Sikhona, pushing us to think bigger and do more. Together, we continue to learn, grow, and strive for excellence.

At Sikhona, we are building something bigger than any one of us. We're building a movement, a family of individuals who are dedicated to making a difference. And as we move forward, we will continue to invest in our people, empowering them to grow both personally and professionally as they help us create the future we all envision.

---

## Building a Legacy of Innovation and Impact

As I sit down to write this, Sikhona is valued at an impressive $2.9 billion—and we're only just beginning. However, I believe this valuation is conservative, especially considering the rapid progress we've made. This figure was determined by a highly reputable venture capital firm, but it only takes into account four of our ten platforms. What's exciting is that we've recently added two more groundbreaking platforms, which are poised to disrupt industries and change the game. With these new additions, I have no doubt that our valuation will continue to soar as our platforms expand and evolve.

Sikhona's growth is nothing short of phenomenal. We're constantly innovating, finding new ways to push boundaries and serve our users more effectively. But let me be clear: while the dollar value is impressive, it's not the number that defines Sikhona. The true measure of our success lies in the real impact we're having on people's lives. Whether it's empowering individuals, creating new opportunities, or fostering stronger communities, the value of Sikhona is felt in the stories of those we've touched.

Looking ahead, my vision for Sikhona is simple yet bold: to continue breaking barriers and leading innovation across multiple sectors. I don't want Sikhona to be a company that follows trends—I want us to set them. We're already positioning ourselves to be leaders in the next frontier of technology, focusing on areas like artificial intelligence, blockchain, and digital inclusion. These aren't just buzzwords to us—they represent opportunities to change the way the world works and to make a lasting, positive impact.

But perhaps most importantly, I want Sikhona to remain anchored in its mission: empowerment, resilience, and community. We are building a company that not only leads in terms of technology but also fosters a sense of belonging and support for those we serve. The road ahead is full of limitless possibilities, and as we continue to innovate, we'll never lose sight of the values that drive us.

---

**More Than a Company**

Sikhona isn't just a company—it's a movement, a declaration, and a legacy. It's proof that innovation doesn't have to come at the expense of values, and that success is most meaningful when it's shared. For me, Sikhona represents everything I've worked for and everything I believe in. It's not just about building platforms—it's about building a better world.

As I reflect on the journey so far, I'm filled with gratitude. Gratitude for the challenges that taught me resilience, for the opportunities that allowed me to grow, and for the people who believed in the vision and helped make it a reality. And while the road ahead is sure to bring new challenges, I'm ready. Because Sikhona isn't just a company—it's a promise. A promise to the world that we are here, we are seen, and we are ready to lead.

# CHAPTER 13
# LEADING WITH PURPOSE – "THE POWER OF VISION AND VALUES"

LEADERSHIP IS ABOUT MORE than just making decisions—it's about defining a vision and aligning people, resources, and opportunities to bring that vision to life. When I look back on my journey, every success and every setback has been tied to one central question: What kind of leader do I want to be? For me, the answer has always been clear: I want to lead with purpose.

Sikhona has given me the platform to do just that. From the beginning, my goal wasn't just to build a successful company—it was to create something meaningful. Something that would stand the test of time, not just because of its financial success, but because of its impact on people's lives. That's the power of leading with vision and values.

---

### Defining the Vision

When I started Sikhona, I had a clear vision: to create a digital ecosystem that empowers individuals and communities to thrive. That meant more than just building platforms

—it meant addressing real-world problems in innovative ways. It meant bridging gaps between cultures, creating opportunities for marginalized voices, and fostering a sense of belonging in an increasingly digital world.

But vision alone isn't enough. A leader has to be able to communicate that vision clearly and passionately, inspiring others to believe in it as much as they do. For me, that's always come naturally. I don't just see Sikhona as a business —I see it as a movement, a declaration of what's possible when we refuse to accept the limitations imposed on us by society.

From the engineers coding our platforms to the users sharing their stories, everyone involved in Sikhona understands that they're part of something bigger. That's the power of a shared vision—it turns a company into a community and a community into a force for change.

---

**The Role of Values**

While vision provides direction, values provide the foundation. At Sikhona, our core values are innovation, resilience, inclusivity, and integrity. These aren't just words on a mission statement—they're principles that guide every decision we make.

Innovation means we're constantly pushing boundaries, looking for new ways to solve problems and create value. Resilience means we don't back down in the face of challenges—we adapt, learn, and grow stronger. Inclusivity means we're committed to creating a platform where everyone feels welcome, regardless of their background, beliefs, or identity. And integrity means we do what's right, even when it's not the easiest or most profitable choice.

These values are woven into the fabric of Sikhona. They inform how we design our platforms, how we interact with our users, and how we make decisions as a team. They're also a reflection of my own journey, shaped by the lessons I've learned along the way.

---

**Lessons in Leadership**

Leadership isn't something you're born with—it's something you develop through experience. Over the years, I've learned that being a leader means making tough decisions, taking responsibility for your actions, and always staying true to your values. It also means recognizing that you don't have all the answers and being willing to listen, learn, and grow.

One of the most important lessons I've learned is the power of authenticity. People don't just follow leaders because of their title or position—they follow them because they believe in them. As a leader, I've always tried to be honest and transparent, even when the truth is difficult. That's how you build trust, and trust is the foundation of any successful team.

Another key lesson is the importance of adaptability. The world is constantly changing, and leaders have to be able to change with it. At Sikhona, we've faced our share of challenges, from technical setbacks to funding hurdles to external criticism. Each one has been an opportunity to learn, adapt, and come back stronger.

Finally, I've learned that leadership is about service. It's not about telling people what to do—it's about empowering them to do their best work. At Sikhona, I strive to create an environment where everyone feels valued, supported, and inspired to contribute to our shared vision.

---

**Leading Through Challenges**

Leadership is often tested in moments of adversity, and Sikhona has faced its share of challenges. One of the toughest decisions I've had to make was turning down a $200 million offer from a venture capital firm. On paper, it seemed like an incredible opportunity. But the terms of the deal would have compromised Sikhona's mission and values, and I wasn't willing to let that happen.

Another major challenge was navigating the backlash we faced during the transition from OBT Social to Sikhona. Some people questioned the rebranding, while others doubted our ability to expand our mission beyond a single community. Those criticisms hurt, but they also motivated me to prove the doubters wrong. By staying true to our vision and values, we were able to turn those challenges into opportunities for growth.

---

**Looking Ahead**

As I think about the future of Sikhona, I'm filled with excitement and optimism. Our platforms are growing rapidly, and we're constantly exploring new ways to innovate and serve our users. From artificial intelligence to blockchain to digital inclusion, the possibilities are endless.

But no matter how much we grow, one thing will never change: our commitment to leading with purpose. Sikhona will always be a company driven by vision and values, focused on creating a better future for everyone.

---

## The Power of Vision and Values

At its core, leadership is about creating change. It's about imagining a better world and working tirelessly to make that vision a reality. That's what Sikhona represents to me—a chance to lead with purpose, to inspire others, and to leave a lasting legacy.

As I reflect on my journey, I'm reminded of something I've always believed: success isn't just about what you achieve—it's about how you achieve it. By staying true to my vision and values, I've been able to create something I'm truly proud of. And as Sikhona continues to grow, I'm excited to see how we can inspire others to do the same.

# CHAPTER 14
# THE FUTURE OF TECHNOLOGY – "WHAT'S NEXT FOR SIKHONA"

**EXPANDING the Ecosystem**

Sikhona's ecosystem is already robust, with platforms like **HeyYo**, **DoYou**, **File Mule**, and **Sikhona University** addressing diverse needs. But this is just the beginning. The future of Sikhona lies in expanding this ecosystem, creating new platforms that address emerging challenges and opportunities.

One area we're exploring is artificial intelligence (AI). AI has the potential to revolutionize industries, from healthcare to education to entertainment. At Sikhona, we're developing AI-driven tools that can personalize user experiences, enhance productivity, and democratize access to technology. Imagine an AI-powered virtual tutor that adapts to a student's learning style or a creative assistant that helps content creators brainstorm and execute their ideas. These are the kinds of solutions we're working on.

Another focus is blockchain technology. Blockchain offers a level of transparency, security, and decentralization that can transform how we interact with digital systems. We're exploring ways to integrate blockchain into Sikhona's platforms, from secure file sharing on File Mule to digital credentialing on Sikhona University. Our goal is to leverage blockchain not just as a buzzword but as a meaningful innovation that adds real value to our users.

### Connecting the Global Community

One of Sikhona's core missions is to connect people across borders. The digital divide is still a significant issue, with millions lacking access to the tools and resources they need to thrive in a digital age. As we grow, we're committed to addressing this gap and ensuring that no one is left behind.

We're investing in initiatives that bring affordable internet access to underserved communities, particularly in Africa and other emerging markets. We're also working on localized versions of our platforms, ensuring they meet the unique needs and preferences of different cultures. Sikhona isn't just about creating technology—it's about creating technology that resonates with people on a personal level.

The global community is also a key focus for our educational efforts. Sikhona University is expanding its offerings to include courses in digital literacy, entrepreneurship, and emerging technologies. By equipping individuals with the skills they need to succeed, we're not just preparing them for the future—we're empowering them to shape it.

### Innovation with Purpose

At Sikhona, innovation isn't just about creating something new—it's about creating something meaningful. Every platform, every feature, and every initiative is guided by the question: How will this improve people's lives? That's the standard we hold ourselves to, and it's what sets us apart from other tech companies.

Take **DoYou**, for example. While other short-form video platforms focus on entertainment, DoYou also emphasizes education and self-expression. It's a place where creators can share knowledge, showcase their talents, and build meaningful connections. Similarly, **File Mule** isn't just a file-sharing service—it's a secure, user-friendly solution that prioritizes privacy and accessibility.

This commitment to purposeful innovation extends to our partnerships as well. We're collaborating with organizations that share our values, from nonprofits working on digital inclusion to tech companies pushing the boundaries of what's possible. Together, we're building an ecosystem that's greater than the sum of its parts.

---

**Navigating Challenges**

The future of technology is full of promise, but it's not without its challenges. Issues like data privacy, misinformation, and ethical AI are pressing concerns that require thoughtful solutions. At Sikhona, we're not shying away from these challenges—we're tackling them head-on.

Data privacy is one of our top priorities. In a world where personal information is often exploited for profit, we're committed to creating platforms that respect and protect our users' data. We're also investing in tools to combat misinformation, ensuring that our platforms remain trusted sources of information and connection.

When it comes to AI, we're taking a proactive approach to ethics. AI has the power to do incredible good, but it also comes with risks. That's why we're establishing guidelines for responsible AI development and partnering with experts to ensure that our tools are fair, transparent, and aligned with our values.

---

**A Legacy of Leadership**

As I think about Sikhona's future, I'm not just focused on the next five or ten years—I'm thinking about the legacy we'll leave for generations to come. Sikhona isn't just about creating technology—it's about creating a better world. It's about using innovation to empower individuals, uplift communities, and drive positive change on a global scale.

This legacy extends beyond the platforms we build. It's about the culture we create, the values we uphold, and the impact we make. It's about proving that a tech company can be both profitable and purposeful, both innovative and inclusive. That's the legacy I want Sikhona to leave behind.

---

**Inspiring the Next Generation**

One of the most rewarding aspects of leading Sikhona has been inspiring the next generation of innovators and leaders. I want young people to see that success doesn't require a traditional path. You don't need a fancy degree or a perfect resume—you just need a vision, a willingness to learn, and the determination to keep going, no matter what.

Through Sikhona University and our other initiatives, we're creating opportunities for young people to discover their passions, develop their skills, and turn their dreams into reality. Whether it's a budding entrepreneur in Lagos, a student in Detroit, or a creator in Mumbai, I want them to know that they have a place in Sikhona's ecosystem.

---

**A Vision for the Future**

As Sikhona continues to grow, our vision remains the same: to be a company that doesn't just follow trends but sets them. To lead with purpose, innovate with integrity, and create a lasting impact on the world. The road ahead will undoubtedly bring new challenges, but I'm confident that we're ready to face them.

Sikhona is more than just a company—it's a movement, a declaration, and a promise. A promise to use technology for good, to empower those who've been overlooked, and to leave the world better than we found it. And as I look to the future, I'm more excited than ever to see where this journey takes us.

# CHAPTER 15
# A LEGACY IN THE MAKING – "BUILDING BEYOND THE PRESENT"

SUCCESS ISN'T JUST about what you achieve today—it's about what you leave behind. For me, Sikhona isn't just a company or even a movement. It's my legacy. It's the culmination of everything I've learned, fought for, and dreamed about over the years. As I think about what comes next, I'm focused on building something that will stand the test of time, not just for me but for the communities and individuals who've been part of this journey.

---

**The Meaning of Legacy**

Legacy is a word that gets thrown around a lot, but for me, it has a very specific meaning. It's about impact—about making a difference in people's lives that lasts long after you're gone. It's about creating something bigger than yourself, something that continues to grow and evolve.

When I think about legacy, I don't just think about financial success or technological achievements. I think about the people whose lives have been changed because of what we've built. I think about the students who gained new skills through Sikhona University, the creators who found their voice on HeyYo, and the communities that came together through our platforms.

Sikhona is my way of saying, "We were here. We made a difference." It's a reminder that innovation and progress can come from anywhere and that everyone deserves a seat at the table.

---

### A Legacy of Empowerment

One of the things I'm most proud of is Sikhona's commitment to empowerment. From the beginning, our mission has been to uplift individuals and communities, particularly those who've been overlooked or marginalized. Whether it's through education, entrepreneurship, or creative expression, we're giving people the tools they need to succeed.

This commitment is especially important to me as someone who didn't take the traditional path to success. I didn't have an Ivy League degree or a Silicon Valley pedigree. What I had was vision, determination, and a willingness to learn. That's what I want Sikhona to represent—the idea that success is possible for anyone, regardless of where they come from or what they've been told.

Through our platforms, we're helping people turn their dreams into reality. We're showing them that they're not just consumers of technology—they're creators, innovators, and leaders. And in doing so, we're creating a legacy that's rooted in empowerment.

---

## The Next Generation of Leaders

Legacy isn't just about what you build—it's about who you inspire. One of the most rewarding aspects of leading Sikhona has been seeing the next generation of leaders emerge. From our team members to our users, I've had the privilege of watching countless individuals step into their own power and realize their potential.

At Sikhona, we're intentional about nurturing talent. We're not just building platforms; we're building people. Whether it's through mentorship programs, leadership opportunities, or professional development initiatives, we're investing in the future. Because the truth is, Sikhona's legacy isn't just about me—it's about the countless people who will carry this vision forward.

---

## A Global Perspective

A legacy that doesn't extend beyond borders isn't a legacy—it's a bubble. That's why Sikhona has always had a global perspective. We're not just focused on one market or one audience. We're building a platform that connects people across cultures, continents, and languages.

This global approach is especially important when it comes to addressing the digital divide. Millions of people around the world still lack access to basic internet services, let alone the tools and resources needed to thrive in a digital age. At Sikhona, we're committed to changing that. We're working to bring affordable internet access to underserved communities, develop localized versions of our platforms, and ensure that our impact is felt everywhere.

By thinking globally, we're not just expanding our reach —we're expanding our legacy. We're showing the world that innovation doesn't have to come at the expense of inclusivity and that progress is most meaningful when it's shared.

---

**Challenges Along the Way**

Building a legacy isn't easy. It requires resilience, adaptability, and a willingness to face challenges head-on. Sikhona has encountered its fair share of obstacles, from funding hurdles to external criticism. But every challenge has been an opportunity to grow, learn, and strengthen our resolve.

One of the most significant challenges has been staying true to our mission in the face of external pressures. I've had offers to sell Sikhona or take on investors who would have steered the company in a different direction. While those offers were tempting, I knew that accepting them would mean compromising the very values that make Sikhona special.

Another challenge has been navigating the complexities of building a global platform. Every market is different, with its own needs, preferences, and obstacles. But by listening to our users, collaborating with local partners, and staying focused on our mission, we've been able to overcome these challenges and continue moving forward.

---

**A Legacy of Innovation**

At its core, Sikhona is a legacy of innovation. It's a testament to the power of ideas and the impact they can have when they're brought to life with passion and purpose. From our early platforms to the groundbreaking work we're doing today, innovation has always been at the heart of what we do.

But innovation isn't just about technology—it's about mindset. It's about challenging the status quo, questioning assumptions, and constantly striving for something better. That's the legacy I want Sikhona to leave behind—a legacy of bold thinking, relentless curiosity, and unwavering commitment to progress.

---

### Looking Ahead

As I think about the future, I'm filled with a sense of excitement and responsibility. Sikhona's story is far from over, and there's so much more I want to accomplish. From expanding our platforms to exploring new frontiers in technology, the possibilities are endless.

But more than anything, I want Sikhona's legacy to be one of impact. I want to know that we made a difference, that we created opportunities for people who needed them, and that we showed the world what's possible when you lead with vision and values.

---

### A Legacy in the Making

Legacy isn't something you achieve—it's something you build, piece by piece, over time. Sikhona is still in its early chapters, but I'm proud of what we've accomplished so far. And as we continue to grow, I'm committed to staying true to the mission, values, and vision that brought us here.

This isn't just about me—it's about everyone who's been part of this journey. It's about the users, the team members, the partners, and the communities who've believed in Sikhona and helped make it what it is today. Together, we're building something that will stand the test of time. Together, we're building a legacy.

---

**Legacy and Reflection**

Now that I've had time to reflect on my journey, I see the patterns more clearly. The kid who took apart his toys to build something new, the teenager who dreamed big despite pushback, the young adult who worked himself to the bone—it's all connected. Every stage of my life has been about creating, innovating, and pushing boundaries.

One of the things I'm most proud of is the legacy I'm building through Sikhona. It's not just about technology or business—it's about leaving something meaningful behind. When I look at the work we're doing, I see the potential to change lives, to create opportunities for people who might not have had them otherwise. That's what drives me now.

I've also learned to value the people who've been part of my journey. Whether it's my family, my friends, or the teams I've worked with, they've all played a role in shaping who I am. And for that, I'm deeply grateful.

Looking back, I wouldn't change a thing. Every challenge, every setback, every victory—it's all part of the story. And the story's not over yet.

Technology has always been a driving force behind change. It shapes how we live, work, and connect with one another. At Sikhona, we see technology not just as a tool but as a platform for empowerment. As we look toward the future, our mission remains clear: to stay ahead of the curve, anticipate the needs of tomorrow, and create solutions that make a lasting impact.

This isn't just about keeping up with trends. It's about defining them. Sikhona isn't content to follow the path laid by others—we aim to blaze our own trail. And as we stand on the cusp of the next wave of innovation, I couldn't be more excited about what lies ahead.

# CHAPTER 16
# DISCUSSION AND REFLECTION QUESTIONS

### CHAPTER 1: **Early Beginnings – "A Curious Mind Unleashed"**

Michael Thompson's early experiences were defined by a deep curiosity and a passion for exploration. From dismantling toys to experimenting with his Tyco racing car set, his childhood was a fertile ground for nurturing creativity and problem-solving. His upbringing in Northern Virginia, paired with an insatiable desire to learn, laid the foundation for a lifetime of innovation.

**Discussion and Reflection Questions**
_____

1. How did Michael's childhood experiences with curiosity and experimentation shape his later successes?

2. In what ways can childhood interests influence career paths? Reflect on your own experiences.

3. Michael took apart toys to learn how they worked. What is an example of something you've dismantled or deeply explored to understand better?

4. How do early influences, like family or environment, play a role in shaping one's ambitions?

5. Michael set a promise for himself early in life. What is a promise or goal you set for yourself as a child, and how has it influenced your life?

---

## Chapter 2: Seeds of Innovation – "Creative Youth"

With limited resources, Michael's creativity thrived. He transformed everyday objects into meaningful projects, such as using a Light Bright to create a makeshift sunlight source for growing watermelon seeds. These moments of ingenuity illustrate how his resourcefulness became a core strength in his journey.

### Discussion and Reflection Questions

---

1. Michael turned limited resources into opportunities for innovation. How can resourcefulness be a strength in challenging situations?

2. What parallels can you draw between Michael's childhood experiments and problem-solving in your own life?

3. How does Michael's approach to creating and learning illustrate the power of persistence? Share a time when persistence paid off for you.

4. Reflect on how childhood hobbies or projects can lead to valuable skills in adulthood. What skills did you unknowingly develop as a child?

5. Michael's creativity was sparked by necessity. How do constraints or limitations encourage innovation in your experience?

## Chapter 3: Shaping Aspirations – "Early Adolescence"

As Michael transitioned into adolescence, he experienced moments of loss, discovery, and ambition. The challenges of his family's modest economic status and the loss of his grandfather shaped his values and aspirations, giving him clarity about what he wanted in life.

### Discussion and Reflection Questions

1. How did Michael's family environment and experiences in Reston shape his understanding of success and community?

2. Loss played a significant role in Michael's early adolescence. How do moments of personal loss shape character and ambition?

3. Michael encountered contrasting life paths among his peers. Reflect on a time when you noticed similar differences in your circle and how it influenced you.

4. What lessons can be learned from Michael's entrepreneurial spirit and determination to challenge traditional paths?

5. How do you balance following your dreams while respecting the advice and concerns of those close to you, as Michael navigated with his father?

---

## Chapter 4: Embracing the Hustle – "High School Years"

High school was a time of entrepreneurial discovery for Michael. Leading a rap group, organizing parties, and managing events honed his leadership skills. These experiences taught him the importance of planning, execution, and adapting to challenges.

### Discussion and Reflection Questions

1. Michael found leadership in unexpected places,

like his rap group and event planning. How have you stepped into leadership roles in unconventional ways?

2. What strategies did Michael use to ensure his events were successful? How can these strategies apply to other types of projects?

3. Reflect on how extracurricular activities or hobbies during your school years shaped your work ethic and skills.

4. Michael's entrepreneurial spirit was evident in his event planning. What is an example of a time you created an opportunity instead of waiting for one?

5. How did Michael's early experiences with organizing and managing events prepare him for future challenges?

---

## Chapter 5: Breaking Boundaries – "Young Adulthood"

Young adulthood marked a period of exploration and growth for Michael. From teaching himself BASIC programming to excelling in his early tech roles, he laid the groundwork for a thriving career in technology. Despite systemic barriers, he persevered with determination and resilience.

## **Discussion and Reflection Questions**

1. Michael's exposure to diverse socioeconomic backgrounds influenced his goals. How have diverse environments shaped your perspective?

2. How did Michael's decision to pursue self-taught learning impact his career? Reflect on the value of alternative paths to education.

3. Discuss a time when you, like Michael, identified and seized an emerging opportunity in your field.

4. Michael learned to advocate for himself early in his career. How can self-advocacy change the trajectory of your professional journey?

5. What does Michael's story teach about the importance of resilience in overcoming systemic barriers like racism?

## **Chapter 6: A Visionary Without a Degree – "Breaking the Mold"**

Michael defied traditional expectations by pursuing self-taught education and certifications instead of a formal degree. His ability to adapt, learn, and advocate for himself set him apart in the tech industry and beyond.

## Discussion and Reflection Questions

1. Michael chose an unconventional path by forgoing a degree. What are the benefits and challenges of non-traditional education?

2. Reflect on a time when you had to learn something independently. How did it shape your confidence and skills?

3. How does Michael's story highlight the importance of adaptability in career success?

4. What lessons can be drawn from Michael's proactive approach to advocating for pay and recognition in the workplace?

5. How do you approach challenges or barriers when pursuing a goal? What can you learn from Michael's persistence?

## Chapter 7: Visionary on the Rise

By his mid-twenties, Michael was recognized as a leader in his field. His work ethic, relationships, and vision propelled him into prominent roles, proving the power of dedication and foresight.

## Discussion and Reflection Questions

1. Michael's leadership style emphasized loyalty

and collaboration. How do you build trust and inspire others in your work?

2. Reflect on how balancing work and personal life can be challenging. How do you ensure both areas thrive?

3. Michael's ability to foresee trends gave him a competitive edge. How do you stay ahead in your industry or area of interest?

4. How can past experiences and relationships influence future opportunities, as seen in Michael's career progression?

5. What steps can you take to maintain resilience when pursuing ambitious goals?

---

## Chapter 8: The First Breakthrough – "Food on the Move"

Michael's first entrepreneurial venture, Food on the Move, exemplified his ability to identify inefficiencies and create innovative solutions. His experiences navigating racism and building a business taught him invaluable lessons about resilience and leadership.

### Discussion and Reflection Questions

---

1. How did Michael's ability to identify inefficiencies contribute to the success of Food on the Move?

2. Discuss a time when you turned a challenge or dismissal into an opportunity. How did it shape your perspective?

3. How can entrepreneurship be a powerful tool for addressing systemic challenges, as demonstrated by Michael's story?

4. Reflect on the role of resilience in overcoming discrimination or other external barriers.

5. What steps can you take to turn a passion or idea into a sustainable venture?

## Chapter 9: Breaking Through Tech Barriers – "StreamNow"

Michael's work on StreamNow highlighted his ability to innovate in a rapidly changing industry. He navigated challenges in funding and competition while maintaining his focus on creating user-centric technology.

## Discussion and Reflection Questions

1. How did Michael's vision for StreamNow demonstrate his understanding of market needs?

2. Discuss the importance of persistence in overcoming funding challenges for startups.

3. What lessons can be learned from Michael's ability to balance innovation with practicality?

4. How do you evaluate the potential of a new technology or idea in your field?

5. Reflect on the role of competition in driving excellence and innovation.

---

### Chapter 10: Building a Legacy – "Sikhona"

Sikhona represents the culmination of Michael's entrepreneurial journey. His vision for a platform that empowers communities reflects his dedication to innovation and inclusivity.

### Discussion and Reflection Questions

---

1. How does Sikhona embody Michael's values and vision for the future?

2. What strategies did Michael use to ensure Sikhona's success in a competitive market?

3. Reflect on the importance of creating products or services that prioritize community empowerment.

4. How can you apply Michael's approach to leadership and vision in your own projects?

5. What steps can you take to leave a meaningful legacy in your personal or professional life?

---

## Chapter 11: The Global Impact of Technology – "A Vision Beyond Borders"

Michael's global perspective allowed him to recognize the transformative power of technology in underrepresented communities. His efforts to bridge digital divides showcased his commitment to equity and inclusivity.

### Discussion and Reflection Questions

1. How did Michael's global perspective shape his approach to technology and innovation?

2. Reflect on the role of technology in addressing inequalities. How can it be leveraged for social good?

3. What challenges arise when trying to implement global solutions for local problems?

4. Discuss a time when you contributed to a cause that had a global or far-reaching impact.

5. How do you balance innovation with ensuring accessibility and inclusivity?

## Chapter 12: Overcoming Adversity – "Lessons in Resilience"

Michael's journey was filled with obstacles, including systemic racism and financial barriers. His ability to learn from setbacks and persevere inspired others to do the same.

### Discussion and Reflection Questions

1. How did Michael turn adversity into an opportunity for growth?

2. Reflect on a personal challenge you overcame and the lessons it taught you.

3. How can leaders support their teams in navigating adversity and fostering resilience?

4. What role does community play in overcoming systemic challenges?

5. How can adversity shape one's leadership style and character?

## Chapter 13: The Future of Innovation – "Pioneering Tomorrow"

Michael's forward-thinking mindset and emphasis on fostering the next generation of innovators demonstrate his dedication to progress. He believed in empowering youth to create lasting change.

## Discussion and Reflection Questions

1. How can mentoring and empowering youth contribute to a brighter future?

2. Reflect on the importance of fostering creativity and curiosity in younger generations.

3. How do you stay ahead of emerging trends and technologies?

4. Discuss a time when you mentored someone. How did it impact both of your journeys?

5. What steps can you take to ensure your work leaves a positive legacy for future generations?

## Chapter 14: A Personal Legacy – "More Than Success"

For Michael, success was about more than financial gain. His focus on community impact, family, and personal growth highlighted the true meaning of legacy.

## Discussion and Reflection Questions

1. How did Michael define success beyond material achievements?

2. Reflect on the legacy you want to leave in your personal and professional life.

3. How can giving back to the community enhance your sense of fulfillment?

4. What lessons from Michael's journey resonate most with your aspirations?

5. How can you integrate personal values into your professional goals?

---

## Chapter 15: Final Reflections – "Inspiring the Next Generation"

Michael's story serves as an inspiration to those seeking to make a difference. His unwavering dedication to innovation and equity continues to influence future leaders.

### Discussion and Reflection Questions

---

1. What aspects of Michael's story inspire you to take action in your own life?

2. How can you use your skills and passions to create a positive impact?

3. Reflect on a time when someone's story motivated you to pursue a goal.

4. What steps can you take to inspire and support others in your community?

5. How can you ensure your actions align with your values and aspirations?

---

# CHAPTER 17
# MAKING A DIFFERENCE TOGETHER

Sikhona's story continues...

*What next? What will we do together?*

I can hardly wait to see.

Visit me at
https://sikhona.social/static/sikhona

Made in the USA
Columbia, SC
12 February 2025